Petroleum Engineering

Sustainable Geoenergy Engineering and Technology

The objective of this series is to establish a link between traditional hydrocarbon science and engineering and the development of sustainable engineering to achieve net-zero and sustainability goals in energy production. The series aims to promote and expedite the dissemination of new research results, innovations, and technology in the field of exploration, production, and management of oil and gas, aiding the oil and gas industry in navigating through the energy transition and driving it towards the net-zero target. Additionally, the series covers developing areas related to the geological subsurface, such as geothermal energy and underground storage. It includes monographs, lecture notes, and edited volumes, with a subject focus on upstream processes. Coverage extends to all theoretical and applied aspects of the field, encompassing both traditional and modern topics, including but not limited to:

Exploration Formation evaluation (well logging) Drilling Economics Reservoir simulation Reservoir engineering Well engineering Artificial lift systems Facilities engineering Sustainable energy exploration and production Carbon capture, utilization, and storage (CCUS) Geothermal energy development Underground Hydrogen Storage (UHS) Information technology (IT)

Contributions to the series can be made by submitting a proposal to the responsible Springer contact, Anthony Doyle atanthony.doyle@springer.com.

Bodhisatwa Hazra · Debanjan Chandra ·
Vikram Vishal

Unconventional Hydrocarbon Reservoirs: Coal and Shale

Springer

Bodhisatwa Hazra
Department of Coal-Rock Chemistry and Structure
Central Institute of Mining and Fuel Research
Dhanbad, Jharkhand, India

Debanjan Chandra
Department of Geoscience and Engineering
Delft University of Technology
Delft, The Netherlands

Vikram Vishal
Department of Earth Sciences
Indian Institute of Technology Bombay
Mumbai, Maharashtra, India

ISSN 2366-2646 ISSN 2366-2654 (electronic)
Petroleum Engineering
ISBN 978-3-031-53483-6 ISBN 978-3-031-53484-3 (eBook)
https://doi.org/10.1007/978-3-031-53484-3

This Springer imprint is published by the registered company Springer Nature Switzerland AG
The registered company address is: Gewerbestrasse 11, 6330 Cham, Switzerland

Paper in this product is recyclable.

Preface

Unconventional hydrocarbon reservoirs, viz. coals and shales, have received considerable research interest in the past decade or so, due to their strategic significance as they have emerged as a vast and cleaner energy option, and providing the dual benefit of sequestrating atmospheric CO_2 emissions due to their storage properties. Consequently, a lot of research has focused on their assessment for hydrocarbon generation, porosity for elucidating gas storage mechanisms, gas flow mechanisms and upscaling, and the mechanism of sequestrating atmospheric CO_2 in them. Thermal maturity level, the type of kerogen present, quantity of organic matter (for shales), pore structural facets, mineralogy, depth, and thickness have been recognized as significant factors that control hydrocarbon producibility and well as storage potential of these unconventional reservoirs. Because of their large resources, cleaner energy option, and storage potential, unconventional shales and coals have been identified as a possible bridge towards our transition to renewable energy resources.

In this context, the presented monograph comprehensively explores unconventional reservoirs, mainly coal and shale, and discusses the complex factors that govern their success. Firstly, the fundamentals are discussed such as the multi-faceted techniques employed by different researchers for characterization of these reservoirs. Drawing from our recent research, insights are offered into the techniques and protocols essential for generating reliable data. Moving forward, petroleum geochemistry tools to delineate source-cum-reservoir rock understanding of the subtle aspects of pore structural features that impact gas storage and their complexities are discussed in a step-by-step manner. Essential factors in laboratory studies, including particle crush-size and degassing time-temperature, are carefully studied, providing a comprehensive guide for researchers and students.

The focus of the monograph then shifts to the critical role of reservoir simulation, which is essential for translating insights gained at the laboratory scale to the reservoir scale. The success of shale gas exploration, whether with or without CO_2 injection, depends on this process. Assumptions and basic principles of simulating gas production from shale reservoirs and what are the key input parameters for building a successful reservoir model are discussed. The refinement of a reservoir model involves aligning simulated results with the production history of the reservoir, a process known as 'history matching'. The reservoir simulator provides the means by which field and laboratory data can be integrated and upscaled into a single geologic/reservoir model in order to evaluate exploration and development strategies. Such upscaling is impossible without having a clear idea about the anisotropy and heterogeneity in shale reservoirs. Presence of pre-existing fractures and regional stress conditions also plays crucial role in constructing the accurate geometry and initial condition of the modelled reservoir. Shale reservoirs are classified as triple porosity and dual permeability model, where the induced fracture, matrix and pores within the matrix play critical role in governing gas flow within the reservoir. Based on the experimentally derived hydrological properties, reservoir simulation can provide long-term efficiency of a specific reservoir in a matter of minutes, which helps operators make quick decision on the economic feasibility of shale gas exploration. The discussion extends to how reservoir simulation informs strategic decisions on hydraulic fracturing and well configuration for maximizing gas recovery.

Continuing ahead, the potential of sequestrating captured CO_2 from atmosphere permanently within unconventional coals and shales is explored. It is fairly well established that when CO_2 is injected in gas bearing reservoirs, it is absorbed while the CH_4 is released in its place. This preferential adsorption of CO_2 over CH_4 is significant, especially considering the additional valuation of the CH_4 released and the permanent sequestration of CO_2 in these reservoirs. Thus, the mechanisms of CH_4 and CO_2 adsorption including modelling in shales and coals, the factors that affect the adsorption, the state of research covered globally, and the impact the processes of adsorption and desorption on the reservoir performance are discussed. The impact of organic richness, thermal maturity, kerogen type, and mineralogy on gas storage mechanics in shales and coals are covered. It is also understood that CO_2 when injected in deep reservoirs transit to supercritical state, which can have significant impacts on the reservoir properties. Therefore, the importance/impact of supercritical CO_2 on coal and shale reservoir properties viz. wettability behaviour, geomechanics, petrophysical properties, etc., is also explored as it can affect the wellbore stability, fracturing efficiency, and sealing capacity of these reservoirs for geological storage, especially when long-term sequestration of CO_2 is considered.

The monograph concludes by highlighting some of the research gaps, challenges, and perspectives, while considering unconventional coals and shales for future energy needs and climate change. It is envisaged that for reducing global anthropogenic CO_2 emissions, sequestration of CO_2 in these reservoirs is extremely vital, in addition to other forms of CO_2 abatement.

Dhanbad, India — Bodhisatwa Hazra
Delft, The Netherlands — Debanjan Chandra
Mumbai, India — Vikram Vishal

Contents

Chapter 1
Introduction

Unconventional hydrocarbon reservoirs are loosely defined as those that cannot be produced with conventional techniques, and sophisticated methods including multi-stage hydraulic fracturing and horizontal drilling are used to produce gas and/or oil from these reservoirs. In general, the volume of gas present in different geological reservoirs has been found to increase from conventional reservoirs to unconventional Coal Bed Methane systems to tight sand and shale gas systems to gas hydrates, although more abundant gas resources bring in more significant environmental integrity into play (Islam 2014). These reservoirs, although are of different types, are individually unique is different aspects and in this book, we focus on unconventional coal and shale together, and do not focus on other unconventional hydrocarbon systems.

Hydrocarbons in coal and shale reservoirs are generated from the organic matter present within them primarily due to thermal maturation, and these generated hydrocarbons are stored or retained within the coal and shale systems through different processes with adsorption being the most significant phenomenon (Moore 2012; Loucks et al. 2009; Zhang et al. 2012). Consequently, numerous studies have been carried out in recent years focussing on the mechanisms of generation of hydrocarbons from these reservoirs and how they are stored within these reservoirs (Kim 1977; Levine et al. 1993; Crosdale et al. 1998; Busch and Gensterblum 2011; Moore 2012; Moore et al. 2014; Heller and Zoback 2014; Psarras et al. 2017; Hazra et al. 2018a, b, 2019a, b; Vishal et al. 2019; Chandra et al. 2022, 2023). Because of their unique ability to adsorb gases, coals and shales are seen as potential targets for sequestrating CO_2 and reduce the CO_2 emissions in the atmosphere for sustainable development of society (Heller and Zoback 2014; Vishal et al. 2018; Wang et al. 2024). This ability of the unconventional coal and shale as potential targets for CO_2 sequestration, in addition to them being a cleaner energy option, is especially significant for offsetting the atmospheric emissions caused by burning fossil fuels, and will eventually make it possible to mitigate the captured CO_2 in a long-lasting and sustainable way

B. Hazra et al., *Unconventional Hydrocarbon Reservoirs: Coal and Shale*, Petroleum Engineering, https://doi.org/10.1007/978-3-031-53484-3_1

(Vishal et al. 2021a, b; Hazra et al. 2022). Sequestration of CO_2 in deep coal and shale reservoirs also presents the additional advantage of enhanced gas recovery from CBM (CO_2-ECBM) and shale gas (CO_2-EGR) reservoirs, which can substantially offset the overall cost of CO_2 storage projects and presents enormous possibilities of reduction of CO_2 with the production of methane, lowering these unconventional reservoirs' carbon footprint, while improving its productivity (Gentzis 2000; Godec et al. 2014).

Shale reservoirs, due to their lack of fluid migration pathways, were long considered unsuitable for storing CO_2. However, the technological advancement in hydraulic fracturing and its application in different reservoirs worldwide has reawakened the possibility for CO_2 storage in shales. Multiple studies were performed, where the possibility of pore volume and permeability increase were explored with chemical injection or thermal stimulations (Chandra et al. 2021, 2023; Hazra et al. 2023; Edgin et al. 2021; Mouzakis et al. 2016). Moreover, injection of supercritical CO_2 itself has been reported to increase shale porosity and permeability, which has a positive impact on long term CO_2 injection in coal and shales (Cheng et al. 2021; Dai et al. 2020; Fatah et al. 2022; Yin et al. 2016). Coal and shale, similar to any sedimentary rocks, are highly heterogeneous and anisotropic depending on the change in depositional conditions (Hazra et al. 2016; Sethi et al. 2024). These heterogeneities are strictly scale-dependent and largely affects the gas extraction and migration behavior of the reservoir. The underlying causes of these heterogeneities and their effects have been demonstrated in multiple studies (Bahadur et al. 2023; Chandra et al. 2022; Fathi and Akkutlu 2009; Gao et al. 2021; Tian et al. 2022).

Characterizing the physical properties of shale and coal are the first step to understand their potential as a reservoir. As pores serve as the main locations for gas storage in shale, understanding the pores and their connectivity is of utmost importance. Many 2D and 3D imaging methods are being developed to describe the pores, fractures and their mutual interaction in shale and coal, which further assisted fluid flow modelling in these reservoir rocks (Bultreys et al. 2016; Chandra and Vishal 2021), however merging information obtained across different scales up to the reservoir scale has always been challenging. Constructing extremely detailed reservoir models are time consuming and using them in numerical simulators used to predict the dynamics of flow in the reservoir can cause multiple issues with model convergence. Often physical properties of the reservoirs are averaged to a higher order to make model construction and simulation time reasonably fast. Modern simulators often incorporate the coupled physics of deformation and flow, which increases the accuracy of the prediction. Both coals and shales are marked by highly heterogeneous complex pore systems, and the CH_4 present in these unconventional reservoirs are adsorbed by the porous structures. Although the nature of gas retention is similar in coal and shale reservoirs, a direct comparison of CH_4 and CO_2 adsorption capacities of coals and shales may not be descriptive mainly due to the fact that the two rock types are different, and consequently the role of their constituting components are different. For CO_2-ECBM and CO_2-EGR as well, the role of the pore systems is extremely significant. Because of their importance and ability to store CH_4 and CO_2, a lot of research in recent years have been directed towards understanding the pore structural complexities of shale and coal reservoirs, employing a variety of methods, such as imaging and scattering methods, high pressure and low pressure gas adsorption, Atomic Force Microscopy, which have been used in recent years to characterize porosity of organic-matter (Mann et al. 1994; Day et al. 2008; Ross and Bustin 2009; Bernard et al. 2012; Weniger et al. 2012; Clarkson et al. 2013; Labani et al. 2013; Yu et al. 2014; Liu et al. 2016, 2017, 2019; Li et al. 2018; Hazra et al. 2018c, 2019b; Valentine et al. 2019; Chandra and Vishal 2020; Chandra et al. 2020; Sethi et al. 2023). In general, for classifying the pore sizes in CBM and shale gas reservoirs, the classification of the International Union of Pure and Applied Chemistry (IUPAC) is followed, which is as follows: micropores (< 2.0 nm in diameter), mesopores (2.0–50.0 nm in diameter), and macropores (> 50.0 nm in diameter) (Rouquerol et al. 1994). In general, thermal maturity levels of the organic matter present within these reservoirs, the type of kerogen present, quantity of organic matter (for shales), pore structural facets, mineralogy, depth and thickness have been recognised as significant factors that controls hydrocarbon producibility and well as storage potential of these unconventional reservoirs. Unconventional shales and coals are being identified as a viable link in the shift to renewable energy sources due to their abundance, cleaner energy alternative, and storage potential.

In this monograph, step-by-step, we provide a detailed treatment of unconventional coal and shale reservoirs. In Chap. 2, we detail the commonly used petroleum geochemistry tool for delineating source-cum-reservoir rocks in terms of their kerogen type, thermal maturity. We lay emphasis on how analytical limitations and fallacies related to over-simplistic interpretations may be avoided. In Chap. 3, we then move to the techniques of determining pore structural facets of unconventional reservoirs, which controls the gas storage potential of these reservoirs. Recent researches have highlighted that while carrying laboratory studies for storage estimates, several critical factors viz. particle crush-size, degassing time–temperature, moisture content, effectiveness of the probe to access the pore structures etc. are to be critically monitored. In a step-by-step manner, we critically cover these points in this section. In Chap. 4 we move to the technique of imaging and visualization of pore systems in coals and shales using different techniques, viz. optical microscopy, electron microscopy, optical-electron correlative microscopy, AFM (atomic force microscopy), neutron and x-ray scattering techniques etc. The success of shale gas exploration with or without CO_2 injection depends on translating laboratory scale experimental understanding discussed in the past few chapters to reservoir scale. Reservoir simulation provides a consistent and reliable way to account for the various complex mechanisms associated with fluid storage and flow in shale reservoirs. In Chap. 5 the assumptions and basic principles of simulating gas production from shale reservoirs and what are the key input parameters for building a successful reservoir model are discussed. We also iterate on how to refine a reservoir model by matching simulated results with production history of the reservoir, traditionally referred as 'history matching'. The reservoir simulator provides the means by which field and laboratory data can be integrated and upscaled into a single geologic/reservoir model in order to evaluate exploration and development strategies. Such upscaling is impossible without having a clear idea about the anisotropy and heterogeneity in shale reservoirs. Presence of pre-existing fractures and regional stress conditions also play crucial role in constructing the accurate geometry and initial condition of the modelled reservoir. Shale reservoirs are classified as triple porosity and dual permeability model, where the induced fracture, matrix and pores within the matrix play critical role in governing gas flow within the reservoir. Based on the experimentally derived hydrological properties, reservoir simulation can provide long-term efficiency of a specific reservoir in a matter of minutes, which helps operators make quick decision on the economic feasibility of shale gas exploration. We also discuss on how reservoir simulation helps strategizing hydraulic fracturing and well (injection/production) configuration to maximize recovery of stored gas. In Chap. 6, we cover the potential of sequestrating captured CO_2 from atmosphere, permanently within unconventional coals and shales. It is fairly well established that when CO_2 is injected in gas bearing reservoirs, it is absorbed while the CH_4 is released in its place. This CO_2 over CH_4 preferential adsorption is significant especially considering the additional valuation of the CH_4 released and the permanent sequestration of CO_2 in these reservoirs. In this chapter we comprehensively cover the mechanisms of CH_4 and CO_2 adsorption in shales and coals, the factors that affect the adsorption, and the state of research covered globally. The impact of organic richness, thermal maturity, kerogen type, mineralogy on gas storage in shales and coals are covered. It is also understood that CO_2 when injected in deep reservoirs transit to supercritical state, which can have significant impacts on the reservoir properties. Consequently, in this chapter we also cover the importance/impact of supercritical CO_2 on coal and shale reservoir properties viz. wettability behaviour, geomechanics, petrophysical properties etc., as it can affect the wellbore stability, fracturing efficiency, and sealing capacity of these reservoirs for geological storage, particularly when long-term or permanent sequestration of CO_2 is being contemplated.

References

Bahadur J, Chandra D, Das A, Vishal V, Agrawal AK, Sen D (2023) Pore anisotropy in shale and its dependence on thermal maturity and organic carbon content: a scanning SAXS study. Int J Coal Geol 273:104268. https://doi.org/10.1016/J.COAL.2023.104268

Bernard S, Horsfield B, Schulz H-M, Wirth R, Schreiber A (2012) Geochemical evolution of organic-rich shales with increasing maturity: a STXM and TEM study of the Posidonia Shale (Lower Toarcian, northern Germany). Mar Pet Geol 31:70–89

Bultreys T, De Boever W, Cnudde V (2016) Imaging and image-based fluid transport modeling at the pore scale in geological materials: a practical introduction to the current state-of-the-art. Earth Sci Rev 155:93–128. https://doi.org/10.1016/j.earscirev.2016.02.001

Busch A, Gensterblum Y (2011) CBM and CO_2-ECBM related sorption processes in coal: a review. Int J Coal Geol 87(2):49–71

Chandra D, Vishal V (2020) A comparison of nano-scale pore attributes of Barakar Formation gas shales from Raniganj and Wardha Basin. India using low pressure sorption and FEG-SEM analysis, J Nat Gas Sci Engn, p 103453

Chandra D, Vishal V (2021) A critical review on pore to continuum scale imaging techniques for enhanced shale gas recovery. Earth Sci Rev 217:103638. https://doi.org/10.1016/J.EARSCIREV.2021.103638

Chandra D, Vishal V, Bahadur J, Sen D (2020) A novel approach to identify accessible and inaccessible pores in gas shales using combined low-pressure sorption and SAXS/SANS analysis. Int J Coal Geol, 103556

Chandra D, Bakshi T, Vishal V (2021) Thermal effect on pore characteristics of shale under inert and oxic environments: Insights on pore evolution. Micropor Mesopor Mater 316:110969. https://doi.org/10.1016/J.MICROMESO.2021.110969

Chandra D, Vishal V, Bahadur J, Agrawal AK, Das A, Hazra B, Sen D (2022) Nano-scale physicochemical attributes and their impact on pore heterogeneity in shale. Fuel 314:123070. https://doi.org/10.1016/J.FUEL.2021.123070

Chandra D, Bakshi T, Bahadur J, Hazra B, Vishal V, Kumar S, Sen D, Singh TN (2023) Pore morphology in thermally-treated shales and its implication on CO_2 storage applications: a gas sorption, SEM, and small-angle scattering study. Fuel 331:125877. https://doi.org/10.1016/J.FUEL.2022.125877

Cheng Y, Zhang X, Lu Z, Pan ZJ, Zeng M, Du X, Xiao S (2021) The effect of subcritical and supercritical CO2 on the pore structure of bituminous coals. J Nat Gas Sci Eng 94:104132. https://doi.org/10.1016/J.JNGSE.2021.104132

Clarkson CR, Solano N, Bustin RM, Bustin AMM, Chalmers GRL, He L, Melnichenko YB, Radliński AP, Blach TP (2013) Pore structure characterization of North American shale gas reservoirs using USANS/SANS, gas adsorption, and mercury intrusion. Fuel 103:606–616

Crosdale PJ, Beamish BB, Valix M (1998) Coalbed methane sorption related to coal composition. Int J Coal Geol 35:147–158

Dai X, Wang M, Wei C, Zhang J, Wang X, Zou M (2020) Factors affecting shale microscopic pore structure variation during interaction with supercritical CO_2. J CO_2 Util 38:194–211. https://doi.org/10.1016/J.JCOU.2020.01.021

Day S, Fry R, Sakurovs R (2008) Swelling of Australian coals in supercritical CO_2. Int J Coal Geol 74:41–52

Edgin MG, Medina-Rodriguez B, Kaszuba JP, Dewey JC, Alvarado V (2021) Geochemical reactions and alteration of pore architecture in saturated shale after injection of stimulation fluid. Fuel 303:120815. https://doi.org/10.1016/J.FUEL.2021.120815

Fatah A, Ben Mahmud H, Bennour Z, Gholami R, Hossain M (2022) The impact of supercritical CO_2 on the pore structure and storage capacity of shales. J Nat Gas Sci Eng 98:104394. https://doi.org/10.1016/J.JNGSE.2021.104394

Fathi E, Akkutlu IY (2009) Matrix heterogeneity effects on gas transport and adsorption in coalbed and shale gas reservoirs. Transp Porous Media 80:281–304. https://doi.org/10.1007/S11242-009-9359-4/METRICS

Gao Q, Cheng Y, Han S, Yan C, Li Y, Han Z (2021) Effect of shale matrix heterogeneity on gas transport during production: a microscopic investigation. J Pet Sci Eng 201:108526. https://doi.org/10.1016/J.PETROL.2021.108526

Gentzis T (2000) Subsurface sequestration of carbon dioxide—an overview from an Alberta (Canada) perspective. Int J Coal Geol 43(1–4):287–305

Godec M, Koperna G, Petrusak R, Oudinot A (2014) Enhanced gas recovery and CO_2 storage in gas shales: a summary review of its status and potential. Energy Proc. 63:5849–5857

Hazra B, Varma AK, Bandopadhyay AK, Chakravarty S, Buragohain J, Samad SK, Prasad AK (2016) FTIR, XRF, XRD and SEM characteristics of Permian shales, India. J Nat Gas Sci Eng 32:239–255

Hazra B, Chandra D, Lahiri S, Vishal V, Sethi C, Pandey JK (2023) Pore evolution during combustion of distinct thermally mature shales: insights into potential in situ conversion. Energy Fuels 37(18):13898–13911

Hazra B, Vishal V, Sethi C, Chandra D (2022) Impact of supercritical CO_2 on shale reservoirs and its implication for CO_2 sequestration. Energy Fuels 36(17):9882–9903

Hazra B, Wood DA, Vishal V, Varma AK, Sakha D, Singh AK (2018a) Porosity controls and fractal disposition of organic-rich Permian shales using low-pressure adsorption techniques. Fuel 220:837–848
Hazra B, Wood DA, Kumar S, Saha S, Dutta S, Kumari P, Singh AK (2018b) Fractal disposition, porosity characterization and relationships to thermal maturity for the Lower Permian Raniganj basin shales, India. J Natural Gas Sci Eng 59:452–465
Hazra B, Wood DA, Vishal V, Singh AK (2018c) Pore-characteristics of distinct thermally mature shales: influence of particle sizes on low pressure CO_2 and N_2 adsorption. Energy Fuels 32(8):8175–8186
Hazra B, Wood DA, Mani D, Singh PK, Singh AK (2019a) Evaluation of shale source rocks and reservoirs, vol 142. Springer International Publishing, Berlin/Heidelberg, Germany
Hazra B, Chandra D, Singh AK, Varma AK, Mani D, Singh PK, Boral P, Buragohain J (2019b) Comparative pore structural attributes and fractal dimensions of Lower Permian organic-matter-bearing sediments of two Indian basins: inferences from nitrogen gas adsorption. Energy Sour Part A: Recov, Utili, Environ Eff 41(24):2975–2988
Heller R, Zoback M (2014) Adsorption of methane and carbon dioxide on gas shale and pure mineral samples. J Unconv Oil Gas Resour 8:14–24. https://doi.org/10.1016/j.juogr.2014.06.001
Holmes R, Aljamaan H, Vishal V, Wilcox J, Kovscek AR (2019) Idealized shale sorption isotherm measurements to determine pore capacity, pore size distribution, and surface area. Energy Fuels 33(2):665–676
Islam MR (2014) Unconventional gas reservoirs: evaluation, appraisal, and development. Elsevier
Kim AG (1977) Estimating methane content of bituminous coalbeds from adsorption data
Labani MM, Rezaee R, Saeedi A, Hinai AA (2013) Evaluation of pore size spectrum of gas shale reservoirs using low pressure nitrogen adsorption, gas expansion and mercury porosimetry: a case study from the Perth and Canning Basins, Western Australia. J Pet Sci Eng 112:7–16
Levine JR, Johnson PW, Beamish BB (1993) High pressure microbalance sorption studies. In: Proceedings 1993 International Coalbed Methane Symposium The University of Alabama, Birmingham, May 17–21, pp 187–196
Li C, Ostadhassan M, Guo S, Gentzis T, Kong L (2018) Application of PeakForce tapping mode of atomic force microscope to characterize nanomechanical properties of organic matter of the Bakken Shale. Fuel 233:894–910
Liu K, Ostadhassan M, Bubach B (2016) Applications of nano-indentation methods to estimate nanoscale mechanical properties of shale reservoir rocks. J Natural Gas Sci Eng 35:1310–1319
Liu K, Ostadhassan M, Zhou J, Gentzis T, Rezaee R (2017) Nanoscale pore structure characterization of the Bakken shale in the USA. Fuel 209:567–578
Liu K, Ostadhassan M, Sun L, Zou J, Yuan Y, Gentzis T, Zhang Y, Carvajal-Ortiz H, Rezaee R (2019) A comprehensive pore structure study of the Bakken Shale with SANS, N_2 adsorption and mercury intrusion. Fuel 245:274–285
Loucks RG, Reed RM, Ruppel SC, Jarvie DM (2009) Morphology, genesis, and distribution of nanometer-scale pores in siliceous mudstones of the Mississippian Barnett Shale. J Sediment Res 79:848–861
Mann U, Neisel JD, Burchard WG, Heinen V, Welte DH (1994) Fluid-rock interfaces as revealed by cryo-scanning electron microscopy. First Break 12:131–136
Moore TA (2012) Coalbed methane: a review. Int J Coal Geol 101:36–81
Moore TA, Bowe M, Nas C (2014) High heat flow effects on a coalbed methane reservoir, East Kalimantan (Borneo), Indonesia. Int J Coal Geol 131:7–31
Mouzakis KM, Navarre-Sitchler AK, Rother G, Bañuelos JL, Wang X, Kaszuba JP, Heath JE, Miller QRS, Alvarado V, McCray JE (2016) Experimental study of porosity changes in shale caprocks exposed to CO_2-saturated brines I: evolution of mineralogy, pore connectivity, pore size distribution, and surface area. Environ Eng Sci 33:725–735. https://doi.org/10.1089/EES.2015.0588/ASSET/IMAGES/LARGE/FIGURE4.JPEG
Psarras P, Holmes R, Vishal V, Wilcox J (2017) Methane and CO_2 adsorption capacities of kerogen in the Eagle Ford shale from molecular simulation. Acc Chem Res 50(8):1818–1828

Ross DJ, Bustin RM (2009) The importance of shale composition and pore structure upon gas storage potential of shale gas reservoirs. Mar Pet Geol 26(6):916–927

Sethi C, Mastalerz M, Hower JC, Hazra B, Singh AK, Vishal V (2023) Using optical-electron correlative microscopy for shales of contrasting thermal maturity. Int J Coal Geol, 104273

Sethi C, Hazra B, Ostadhassan M, Motra HB, Dutta A, Pandey JK, Kumar S (2024) Depositional environmental controls on mechanical stratigraphy of Barakar Shales in Rajmahal Basin, India. Int J Coal Geol 285:104477

Tian J, Liu J, Elsworth D, Leong YK, Li W, Zeng J (2022) Shale gas production from reservoirs with hierarchical multiscale structural heterogeneities. J Pet Sci Eng 208:109380. https://doi.org/10.1016/J.PETROL.2021.109380

Vishal V, Chandra D, Bahadur J, Sen D, Hazra B, Mahanta B, Mani D (2019) Interpreting pore dimensions in gas shales using a combination of SEM imaging, small-angle neutron scattering, and low-pressure gas adsorption. Energy Fuels 33(6):4835–4848

Vishal V, Chandra D, Singh U, Verma Y (2021a) Understanding initial opportunities and key challenges for CCUS deployment in India at scale. Resour Conserv Recycl 175:105829

Vishal V, Mahanta B, Pradhan SP, Singh TN, Ranjith PG (2018) Simulation of CO_2 enhanced coalbed methane recovery in Jharia coalfields, India. Energy 159:1185–1194

Vishal V, Verma Y, Chandra D, Ashok D (2021b) A systematic capacity assessment and classification of geologic CO_2 storage systems in India. Int J Greenhouse Gas Control 111:103458

Wang C, Liu B, Mohammadi MR, Fu L, Fattahi E, Motra HB, Hazra B, Hemmati-Sarapardeh A, Ostadhassan M (2024) Integrating experimental study and intelligent modeling of pore evolution in the Bakken during simulated thermal progression for CO_2 storage goals. Appl Ener 359:122693

Weniger P, Francû J, Hemza P, Krooss BM (2012) Investigations on the methane and carbon dioxide sorption capacity of coals from the SW Upper Silesian Coal Basin, Czech Republic. Int J Coal Geol 93:23–39

Yin H, Zhou J, Jiang Y, Xian X, Liu Q (2016) Physical and structural changes in shale associated with supercritical CO_2 exposure. Fuel 184:289–303. https://doi.org/10.1016/J.FUEL.2016.07.028

Yu H, Jing R, Wang P, Chen L, Yang Y (2014) Preferential adsorption behaviour of CH_4 and CO_2 on high-rank coal from Qinshui Basin, China. Int J Min Sci Technol 24(4):491–497

Zhang T, Ellis GS, Ruppel SC, Milliken K, Yang R (2012) Effect of organic-matter type and thermal maturity on methane adsorption in shale-gas systems 47:120–131

Chapter 2
Source-Rock Geochemistry of Unconventional Plays

Laboratory geochemical screening using appropriate techniques for evaluating hydrocarbon generation potential, thermal maturity, organic richness, kerogen type, and presence of free hydrocarbons, has emerged as a favoured tool for researchers (focusing on basic research) and petroleum industry (for delineation of potential horizons) (Espitalié 1986; Jarvie 1991; Peters and Cassa 1994; Lafargue et al. 1998). The techniques of laboratory geochemical screening which were fundamentally introduced for conventional petroleum systems, over the last few decades have seen extended application for unconventional petroleum systems (Jarvie 2012a, b; Hazra et al. 2017, 2018a, b). Open-system programmed pyrolysis-oxidation techniques and organic petrology have been traditionally used for identifying thermal maturity, kerogen type and organic richness of unconventional shales and coals (Romero-Sarmiento et al. 2016; Inan et al. 2017; Wood and Hazra 2018). While application of organic petrology, generates extremely reliable and critical information, it is tedious in its own and to an extent depends on the expertise of the petrographer. Open-system programmed pyrolysis-oxidation techniques on the other hand, is more simplistic, and generates set of information than can be beneficial for screening a larger suite of samples (Hazra et al. 2019a, b).

Amongst the different instruments used for open-system programmed pyrolysis-oxidation the most preferred options among geochemists seen for characterizing source rocks are, the Rock-Eval device, Hawk Pyrolysis and TOC instruments (Varma et al. 2015a, b, c; Romero-Sarmiento et al. 2016; Li et al. 2020). In general, the data produced from these bulk geochemical screening techniques are easy to interpret, significant, and the methodology is simplistic. Different parameters viz. free hydrocarbon content, heavier hydrocarbon content, pyrolyzable carbon (PC), residual carbon (RC), total organic carbon (TOC), thermal maturity, hydrogen and oxygen indices, are measured using these techniques, which helps is better estimation and delineation of potential plays (Lafargue et al. 1998; Behar et al. 2001).

B. Hazra et al., *Unconventional Hydrocarbon Reservoirs: Coal and Shale*, Petroleum Engineering, https://doi.org/10.1007/978-3-031-53484-3_2

While the above geochemical screening techniques are generally uncomplicated, recent researches have distinctly described how analytical flaws or lack of awareness, and following generalized protocols can result in generation of imprecise data, leading to improper interpretations (Sykes and Snowdon 2002; Dembicki 2009; Carjaval-Ortiz and Gentzis 2015; Hazra et al. 2017, 2019a, b, 2021a, b, 2022a, b). Different factors viz. sample crush-size, weight, kerogen type, organic richness, mineral matter etc. impact the data generated, and induce false interpretations. With these corrections, new revised protocols have been proposed for effective clean characterization of potential hydrocarbon plays. In this chapter, we summarize the recent advances made in terms developing protocols for clean laboratory characterization of unconventional coal and shale reservoirs.

2.1 Basic Functionality of Rock-Eval

In general, Rock-Eval method comprises of two stages (Fig. 2.1): (1) pyrolysis stage; (2) oxidation stage. The ideal analysis sequence when using the Rock-Eval Basic/Bulk-Rock method is as follows: (i) at the beginning of the pyrolysis cycle, the initial isothermal phase lasts for three minutes at 300 °C (inert atmosphere). In this stage, the free hydrocarbons in the samples are released, and the Flame Ionisation Detector (FID) picks them up which is recorded as S1 curve; (ii) after the isothermal phase (S1 curve generation), non-isothermal heating phase starts from 300 to 650 °C at a rate of 25 °C/minute (Behar et al. 2001). In this phase, the heavier hydrocarbons in the kerogen cracks, resulting in the release of gases that is again picked up by FID but is now represented as S2 curve in the recorded data. Additionally, during the S2 phase (up to 400 °C), the oxygenated compounds within the kerogen undergo cracking, releasing CO_2 and carbon monoxide (CO) which are picked up by the infrared (IR) detector and records them in the form of S3 curve. The amount of hydrocarbons (recorded under S1 and S2 curves) and oxygenated compounds from organic matter (recorded under S3 curve), is utilized for the calculation of pyrolyzable carbon (PC); (iii) After reaching 650 °C, the temperature of pyrolysis oven cools down to 300 °C. The sample-loaded crucibles are then transferred to the oxidation chamber, where a subsequent heating ramp from 300 °C in the oxygen atmosphere to an ultimate temperature of 750 or 850 °C is applied (at the heating rate of 20 °C/min). The final oxidation temperature is decided on the basis of whether or not carbonate minerals present in the sample need to be quantified (Hazra et al. 2017, 2019a, b). Decomposition of carbonate minerals mostly occur at temperature beyond 650 °C which is denoted by the S5 peak (Behar et al. 2001). During the oxidation stage, the pyrolyzed kerogen in the sample burns, producing CO and CO_2, which is picked up by IR detector. Ideally, the organic matter undergoes decomposition during the oxidation stage

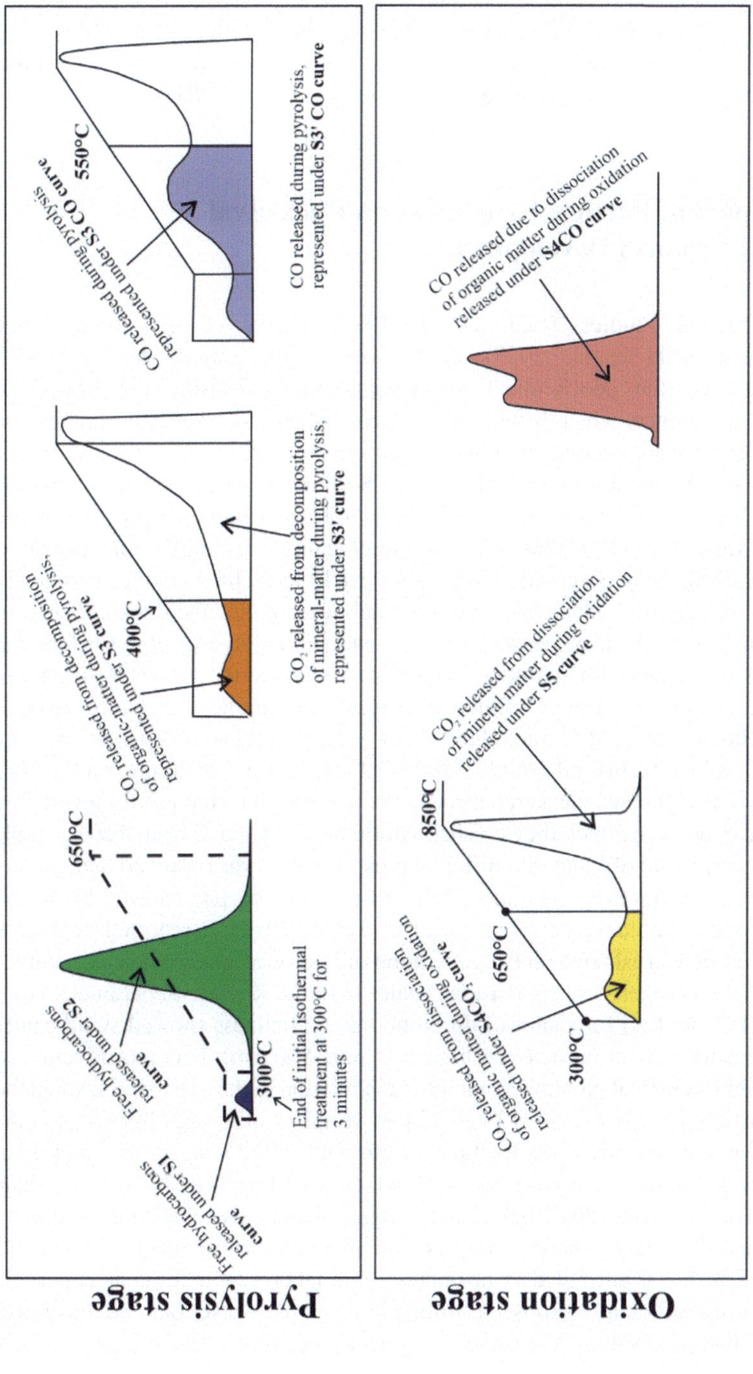

Fig. 2.1 Schematic representation of S1 to S5 peaks produced during Rock-Eval analysis following the basic method (modified after Behar et al. 2001)

at temperatures lower than 650 °C (Behar et al. 2001). The produced CO and CO_2 during oxidation is recorded as S4CO and S4CO_2, respectively. The residual carbon (RC) present in the sample is recorded as S4 curve, which is used for the calculation of TOC of the rock/soil sample (TOC; TOC = PC + RC) (Lafargue et al. 1998; Behar et al. 2001).

2.2 Impact of Particle Crush Size on Rock-Eval S2 and Protocol Development

As per several studies (Inan et al. 1998; Hazra et al. 2017; Singh et al. 2022; Sethi et al. 2023), crush size of the sample has a significant impact on the geochemical parameters generated using the Rock-Eval technique. In essence, when different grain sizes are employed in experiments, the amount of surface area that the number of grains in the samples provides also varies, leading to the generation of variable results. Surface area is crucial in regulating how quickly heat transfer and reactions in the sample can occur during pyrolysis and oxidation. Inan et al. (1998) investigated role of sample size on the data produced from Rock-Eval. They observed that, across particle sizes, the nature of kerogen in humic and organic-rich shales is what ultimately determines the degree of hydrocarbon generation. The least resistance to the generation of hydrocarbons is found in coarser-sized samples with hydrogen-rich kerogen, whereas the greatest resistance to the generation of hydrocarbons is found in coarser-sized samples with hydrogen-poor kerogen. Hazra et al. (2017) studied the effect of grain sizes (< 500 μm, < 250 μm, < 105 μm, and < 63 μm) and sample weight (10 mg, 15 mg, and 20 mg) on the TOC of organic rich sediments measured through the Rock-Eval technique. To investigate the sample crush-size effect, they crushed vitrain band separated from a coal sample from Raniganj Basin of India into different crush-sizes. It was observed that, among the different size fractions, vitrain sample crushed to < 63 μm showed higher S2 across different sample weights (Fig. 2.2). Hazra et al. (2019a, b) reported significant impact of particle crush-size on the generation of hydrocarbons from shale samples for type III-IV kerogen-bearing Permian shales in India. Rock-Eval parameters (S1, S2, TOC, PC and RC) for shales of different organic richness showed significantly lower values for coarser particle crush-sizes (1 mm–500 μm). In contrast, values of the recorded Rock-Eval parameters increased in relation to the crush-sizes, when the sample particle were between 212 and 75 μm. Singh et al. (2022) investigated the effect of particle size (two size fractions: 1 mm and < 212 μm) on the Rock-Eval parameters generated for high-TOC (> 10 wt. %) and low-TOC (< 5wt. %) shale samples. They observed that Rock-Eval pyrolysis of coarse-sized samples recorded lower S2 and PC values, whereas samples with finer particle size recorded higher S2 and PC values. It was attributed to the organic material present in the coarser particle sizes not being sufficiently exposed resulting in incomplete reactions and thus lower S2 and PC recorded values. Conversely, a greater proportion of the organic surfaces

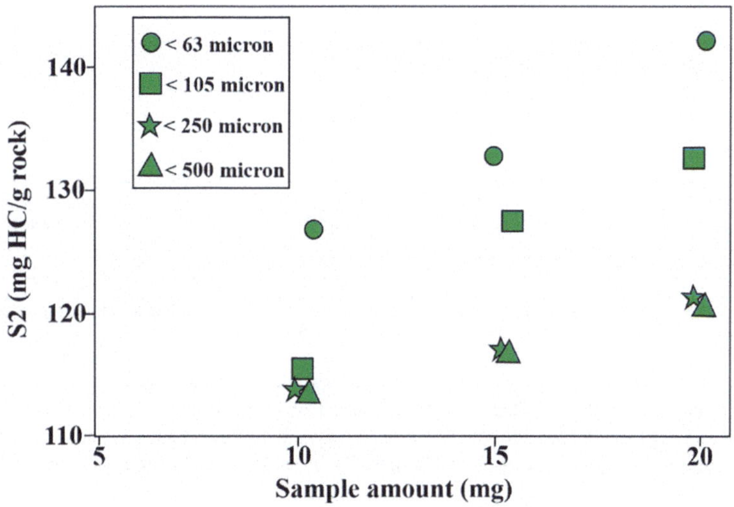

Fig. 2.2 S2 (mg HC/g rock) versus sample amount (mg) plot of the vitrain sample used by Hazra et al. (2017) at various particle crush-sizes and weight. Note that the S2 of finer crush-sizes is higher across all sample weights. Reprinted from International Journal of Coal Geology, Vol. 169, Hazra et al. (2017), TOC calculation of organic matter rich sediments using Rock-Eval pyrolysis: Critical consideration and insights, Pages No. 106–115, Copyright (2017), with permission from Elsevier

become exposed in samples with finer particle sizes, and as a result, they react more thoroughly during pyrolysis, producing higher signals that result in higher S2 and PC recorded values. They noted that the impact of particle size was stronger on the outcomes of Rock-Eval analysis of samples with high TOC. Sethi et al. (2023) made comparable observations in their experimental study to develop protocols for reliable determination of organic geochemical data for shales. Figure 2.3 shows the S2 pyrograms generated from the Rock-Eval pyrolysis of shale samples of different particle sizes (1 mm and < 212 μm) used in their study. It can be observed that samples with particle-crush-sizes of < 212 μm had a higher FID signal and thus a higher S2 curve magnitude than the sample with particle-crush-sizes of 1 mm. Table 2.1 shows the Rock-Eval parameters recorded for samples at different particle crush-sizes. To ensure a reliable comparison of results across various samples, it is crucial to maintain consistency in the crush sizes of the samples during the analysis of the pyrolysis cycle (S1, S2, and S3) (Sethi et al. 2023).

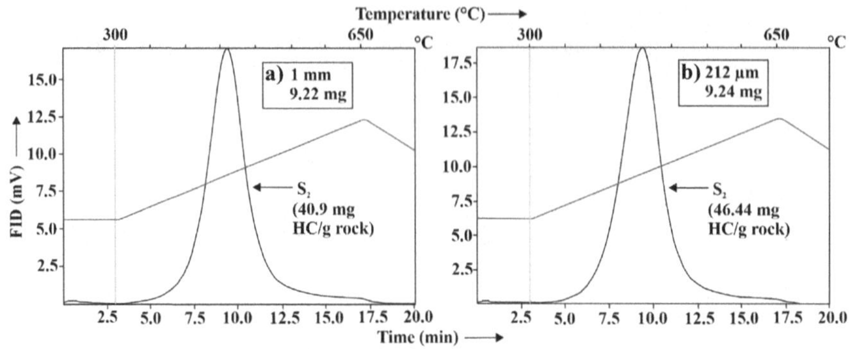

Fig. 2.3 S2 pyrograms of shale samples of two particle sizes: **a** 1 mm particle-crush-size; **b** 212 μm particle crush size. Reprinted from Journal of Earth System Science, Vol. 132, Sethi et al. (2023), Experimental protocols to determine reliable organic geochemistry and geomechanical screening criteria for shales, Copyright (2023), with permission from Springer Nature

2.3 Influence of Sample Weight on Rock-Eval Data

The accuracy of the data obtained from Rock-Eval analysis is significantly affected by sample weight. The Rock-Eval user manual suggests using heavier sample weights for samples that yield a high amount of pyrolysate. Larger samples, however, result in generation of increased amounts of hydrocarbons that can exceed the reliable detection limit of the FID. Thus, instead of having an ideal, narrow Gaussian shape, the S2 curves that are produced represent a broader shape. Carvajal-Ortiz and Gentzis (2015) explained the issue of sample weight for precise geochemical screening using type I-II kerogen-bearing Green River Shale. They observed FID signal reaching 600 mv mark when about 60 mgs of sample was used, which was exceeding the reliable detection threshold of the FID. This resulted in wider S2 pyrogram peak. Conversely, using lower sample weights (~5 mg) for the same analysis, they observed that the produced S2 pyrograms had narrow Gaussian shape and the FID signals was within the reliable detection limit. Additionally, Carvajal-Ortiz and Gentzis (2015) noted that with low sample weights the S2, T_{max} and TOC of the analysed sample also improved in accuracy. So, for samples that yield a lot of oil (type I-II kerogen bearing sedimentary rock samples), they advised using sample weights of about 5 mgs. Hazra et al. (2017) also investigated the sample weight's impact on the values recorded from Rock-Eval analysis using carbonaceous shales and type III-IV organic matter-bearing coal. They observed that the TOC content of the samples decreased with the use of heavier sample weights. According to their findings, the entire organic matter is completely oxidised at lower sample weights below 650 °C (the S4-S5 boundary), but when the sample weights were higher, the organic matter is oxidised over a broader range of temperatures (beyond 650 °C), leading to the release of CO_2 over a wider range of temperatures. As a result, the RC and TOC were ultimately undercounted.

Table 2.1 Rock-Eval recorded parameters of sample of different particle crush-sizes from the literature. Note that the Rock-Eval parameters of finer crush-sizes is higher compared to that of the coarser-sizes

Crush-size	S1 (mg HC/g rock)	S2 (mg HC/g rock)	S3 (mg CO_2/g rock)	T_{max} (°C)	HI (mg HC/g TOC)	OI (mg CO_2/g TOC)	PC (Wt %)	RC (Wt %)	TOC (Wt %)	Sample detail	Literature
1 mm	0.20	40.90	4.33	423	156	17	3.75	22.44	26.21	Permian shale from Rajmahal Basin, India	Sethi et al. (2023)
212 µm	0.24	46.44	4.94	422	170	18	4.23	23.16	27.39		
1 mm	0.13	1.46	0.20	468	72	10	0.18	1.85	2.03	Shale sample from Jharia Basin, India (Lower Permian)	Singh et al. (2022)
212 µm	0.17	1.54	0.54	465	59	21	0.20	2.40	2.60		
1 mm	0.3	44.81	0.12	457	197	1	3.80	18.96	22.76		
212 µm	1.14	8.78	0.21	452	210	1	6.71	30.84	37.55		
1 mm	0.02	2.25	0.27	433	64	8	0.22	3.28	3.50	Shale samples from South Rewa Basin, India (Lower Permian)	
212 µm	0.04	2.68	0.18	431	57	4	0.26	4.44	4.70		
1 mm	0.2	40.42	1.20	423	221	7	3.52	14.77	18.29		
212 µm	0.57	68.61	1.69	429	214	5	5.98	26.11	32.09		
1 mm	0.2	15.07	15.07	424	336	56	1.41	3.07	4.48	Shale samples from Bikaner-Nagpur Basin, India (Palaeocene)	
212 µm	0.18	15.23	15.23	425	311	55	1.43	3.46	4.89		
1 mm	0.85	60.19	5.55	424	409	38	5.45	9.26	14.71		
212 µm	1.24	72.21	5.36	422	438	32	6.48	10.02	16.50		

(continued)

Table 2.1 (continued)

Crush-size	S1 (mg HC/g rock)	S2 (mg HC/g rock)	S3 (mg CO_2/g rock)	T_{max} (°C)	HI (mg HC/g TOC)	OI (mg CO_2/g TOC)	PC (Wt %)	RC (Wt %)	TOC (Wt %)	Sample detail	Literature
1 mm-500 μm	0.22	6.55	1.11	434	111	19	0.64	5.28	5.92	High-TOC shale from Barakar Formation (Lower Permian)	Hazra et al. (2019b)
212–75 μm	0.31	10.76	1.69	434	121	19	1.04	7.85	8.89		
1 mm-500 μm	0.85	32.09	0.40	452	136	2	2.82	20.70	23.52	Carbonaceous shale from Barakar Formation (Lower Permian)	
212–75 μm	1.91	46.85	0.93	451	161	3	4.16	24.94	29.10		
1 mm-500 μm	0.49	60.71	16.84	430	101	28	6.39	53.95	60.34	Vitrain band separated from coal obtained from Barakar Formation (Lower Permian)	
212–75 μm	0.86	73.09	16.23	427	120	27	7.54	53.16	60.70		
< 63 μm	3.65	126.87	6.91	432	163	9	12.00	65.83	77.83	Vitrain band separated from coal obtained from Barakar Formation (Lower Permian)	Hazra et al. (2017)
< 105 μm	2.13	115.4	6.58	432	149	8	10.86	66.58	77.44		
< 250 μm	1.71	114.09	6.61	427	144	8	10.67	68.42	79.09		
< 500 μm	1.49	113.86	7.15	423	147	9	10.45	66.83	77.28		

Table 2.2 displays Rock-Eval data of two coal samples of contrasting thermal maturities from Raniganj basin (RnB) and Jharia basin (JhB) examined with incrementally higher sample weights and crushed to size of 212 μm. Figure 2.4 shows the $S4CO_2$ curves for the two coal samples at the two different sample weights. A closer look at the $S4CO_2$ curves reveals interesting characteristics that relate to the degree of thermal maturity of the coal samples. It can be observed that, at both sample weights, the $S4CO_2$ curves of RnB coal sample (Fig. 2.4a, b) are shifted to the left of the S4-S5 boundary (shown in green in Fig. 2.4a–d). As a result, at higher sample weights, there was no evidence of a decrease in the TOC of RnB coal (Table 2.2). In contrast, even with a smaller sample weight of 5 mgs, the $S4CO_2$ curves for the more thermally mature JhB coal were observed to have small portions located outside the S4-S5 boundary (shown in grey shading in Fig. 2.4c, d). Due to its higher thermal maturity, the organic matter present in JhB coal is more aromatized. This suggests that the organic matter did not experience full combustion at temperatures below 650 °C. As a result, a portion of the $S4CO_2$ curve for JhB coal goes beyond 650 °C (shown in grey shading in Fig. 2.4c, d), and this area of the S4 curve should be considered for calculating RC and TOC. Thus, it is advisable to utilize the entire area under the $S4CO_2$ curve for calculating the RC value to prevent underestimation of a sample's TOC (Hazra et al. 2017, 2019a). The results of coal samples RnB and JhB underscore how the sample weight employed in Rock-Eval analysis and the sample's organic matter thermal maturity impact the accuracy of the generated data. Several other studies have also demonstrated the sample weight's strong influence on the geochemical parameters generated using the Rock-Eval technique (Hazra et al. 2017, 2019a, b, 2020; Singh et al. 2022; Sethi et al. 2023). Typically, it is advisable that higher sample weights may be acceptable for immature shales/coal. Nevertheless, in the case of hydrogen-rich kerogen, elevated levels of hydrocarbons produced during pyrolysis may oversaturate the Flame Ionization Detector (FID) and compromise the reliability of the data (Carvajal-Ortiz and Gentzis 2015). Likewise, in the case of type III-IV kerogen-bearing mature shales/coals, maintaining lower sample weights is recommended to enhance the precision of the generated data.

Table 2.2 Results of Rock-Eval analysis of Raniganj and Jharia basin coals at two sample weights

Basin	SN	Weight	S1 (mg HC/g rock)	S2 (mg HC/g rock)	T_{max} (°C)	HI (mg HC/g rock)	PC (wt %)	RC (wt %)	TOC (wt %)
Raniganj	RnB	4.82	2.08	114.99	425	192	10.56	49.42	59.98
		10.44	2.18	124.31	424	210	11.25	47.92	59.17
Jharia	JhB	4.88	0.94	66.73	463	107	5.82	56.48	62.3
		10.22	0.99	70.85	464	119	6.08	53.48	59.56

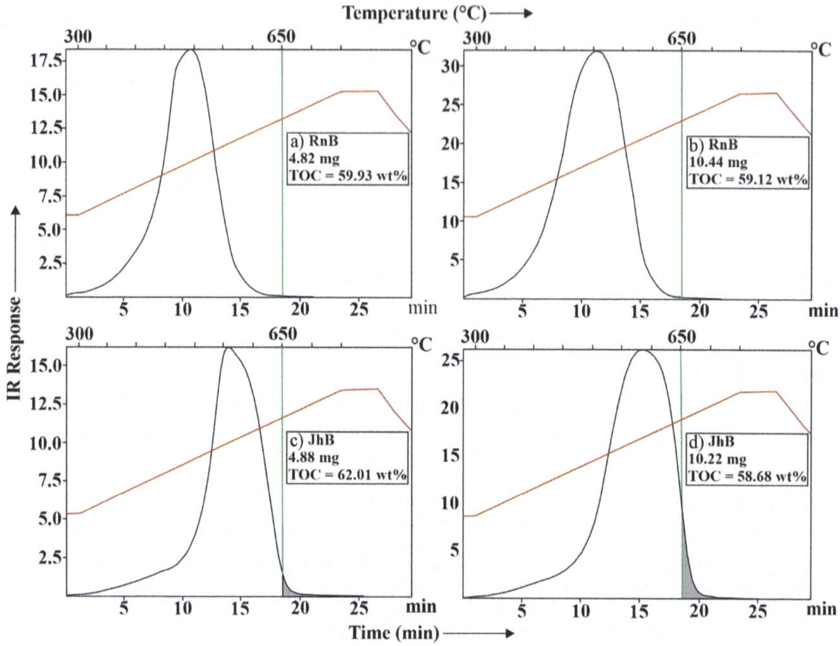

Fig. 2.4 S4CO_2 curves of Raniganj and Jharia basin coals at two sample weights. The portions of the S4CO_2 curves are undercounted in terms of their contribution towards the calculation of TOC

2.4 Impact of Siderite on Rock-Eval S3

The CO and CO_2 liberated through organic matter decomposition during Rock-Eval pyrolysis stage, is represented through S3CO and S3 curves, and data. It is ideally considered that the oxygenated compounds present within organic matter decompose below 400 °C to generate CO and CO_2 (i.e., organic CO and CO_2) during pyrolysis stage. Only the CO2 released under the S3 curve is considered for calculating the oxygen index [OI; OI = (S3/TOC)*100]. Ideally, the OI decreases withing rank of coals and thermal maturities of shales (Peters and Cassa 1994), and hence has been used as proxy for understanding thermal maturity levels of sedimentary organic matter, and is often considered equivalent to the classical O/C ratio of van Krevelen (Peters and Cassa 1994). However, time and again, the reliability of OI and the influence of specific carbonate minerals on S3 and OI has been raised by petroleum geochemists (Katz 1983; Hazra et al. 2022a, b).

Katz (1983) for a suite of carbonate-bearing (calcite and dolomite) rocks, observed very high OI and S3. However, for the same samples subjected to treatment with HCl to eliminate carbonates, there was a substantial decrease observed in both S3 and OI, indicating the role of carbonate minerals on the above-mentioned parameters. Very recently, Hazra et al. (2022a, b) documented that presence of siderite often results in elevated S3 and OI of organic sedimentary rocks. What this essentially means is that carbonates start decomposing at temperatures below 400 °C the CO_2 released from their decomposition overlaps with the CO_2 generated from organic matter decomposition, and thereby resulting in elevated S3 and OI. In nutshell, elevated OI could be due to rank or thermal maturity of the horizon being studied, and also due to presence of certain carbonates which decompose at lower pyrolysis temperatures. Consequently, Hazra et al. (2022a, b) emphasized the significance of examining S3 curves, especially for samples with higher OI, to avoid erroneous thermal maturity interpretations. As mentioned previously, often the simplistic data generated from Rock-Eval may lead to false interpretations. The increase in OI of some shales with higher thermal maturity is a classic example where the increase can be attributed to the presence of carbonates. Consequently, it becomes extremely important for the analysts to cross-check the data generated from analysis, with the curves generated.

Figure 2.5 plots OI versus HI for the Permian shales and coals from India. Published results (200 samples) of Rock-Eval analysis of shales and coals from Lower Permian and Upper Permian Formations from Damodar Valley master-basin, India were collected from literature and used in this figure (Varma et al. 2015a, b, c; Hazra et al. 2015, 2020, 2021a; Mani et al. 2015; Boruah and Ganapathi 2015; Singh et al. 2021). Similarly, published results (129 samples) from Rock-Eval analysis of organic-matter rich rocks from several Indian basins belonging to Cenozoic era were used for comparative analysis (Fig. 2.6; Dutta et al. 2011; Dutta and Chattopadhyay 2014; Paul et al. 2015; Singh et al. 2017a, b; Mathews et al. 2018a, b, 2020; Singh and Kumar 2020; Singh et al. 2020).

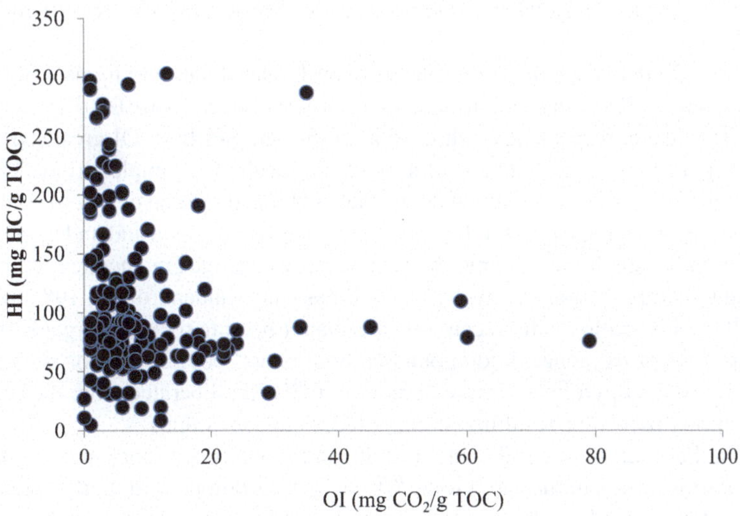

Fig. 2.5 Variation in HI and OI of Permian shales and coals from India (data compiled from Varma et al. 2015a, b, c; Hazra et al. 2015, 2020, 2021a; Mani et al. 2015; Boruah and Ganapathi 2015; Singh et al. 2021)

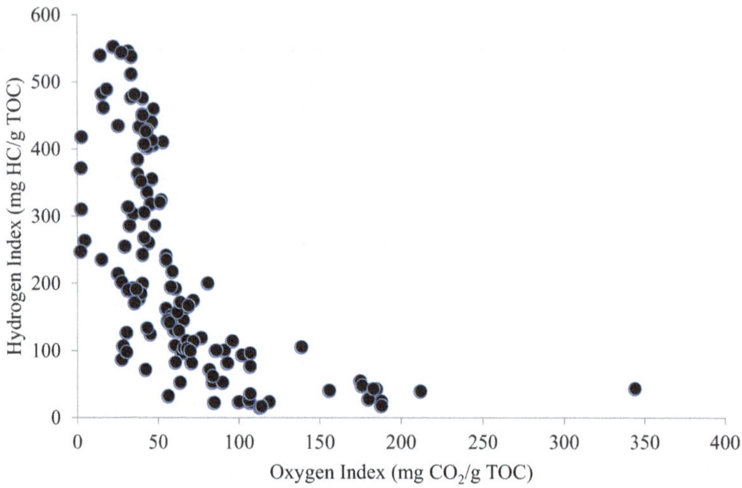

Fig. 2.6 Variation in HI and OI of Cenozoic shales and lignites from India (data compiled from Dutta et al. 2011; Chattopadhyay and Dutta 2014; Paul et al. 2015; Singh et al. 2017a, b; Mathews et al. 2018a, b, 2020; Singh and Kumar 2020; Singh et al. 2020)

Amongst the 200 data collected from literature for organic-rich rocks of Permian Period, for ~ 76% of the samples were observed to have OI $\leq$ 10 mg CO_2/g TOC, while for ~ 16% of the samples, the OI was noted to be varying between 11 and 20 mg CO_2/g TOC. Only 4% of the samples were observed to have OI > 30 mg CO_2/ g TOC.

For the 129 data collected from literature on Indian shales and lignites of Cenozoic Era, only ~ 4% of the samples had OI ranging between 0 to 10 and 11 to 20 mg CO_2/g TOC each, respectively, while 85% of the samples have OI greater than 30 mg CO_2/g TOC (Fig. 2.6). The summary of the results for samples from Permian Period and Cenozoic Era strongly indicate that geological age and degree of coalification has a strong influence on OI. It is reasonably well established and understandable that with increasing rank, organic matter undergoes changes in chemical composition and structure (Tissot and Welte 1978; Behar and Vandenbroucke 1987; Taylor et al. 1998). For geologically recent organic-matter bearing rocks, as organic matter contains a lot of oxygenated compounds (Dutta et al. 2009, 2011), for such rocks during Rock-Eval pyrolysis elevated amount of CO_2 are liberated from the organic matter during pyrolysis, resulting in increased S3 and OI values.

As the Permian shales and coals in India and globally are more mature, the OI in general for these samples are lower. While the variation in OI initially seems less significant, the data does show some inconsistency. For example, two shale samples from Icchapur area of Raniganj basin as reported by Varma et al. (2015a) showed OI values of 91 and 79 mg CO_2/g TOC, while the other shales from the same area showed lower OI. High OI for only few samples amongst a larger set of Permian Indian samples can also be observed from the data of Hazra et al. (2020) and Singh et al. (2021). As the number of samples showing higher OI was very small, the reasons for the same were not explored/documented. Closer inspection of the data-set shows that for samples where OI is higher, the S3' is also high with respect to its S3 and TOC. S3 and S3' curves of few samples where both show stronger IR response is shown in Fig. 2.7.

Kotarba et al. (2002) in their work on Upper Carboniferous coals and shales, Poland, observed ≤ 10 mg CO_2/g TOC OI for majority of the samples, while few samples showed higher values (Fig. 2.8). One sample showed OI of 208 mg CO_2/g TOC, and they inferred probable role of mineral decomposition during pyrolysis to have caused such anomalously high value. However, couple of other samples, from the same suite showed OI as high as 50 mg CO_2/g TOC, although the reasons for the same were not explored. Considering the above results of 28 samples, and the inference by Kotarba et al. (2002), higher OI in some samples could be related to the decomposition of certain carbonate minerals such as siderite during pyrolysis.

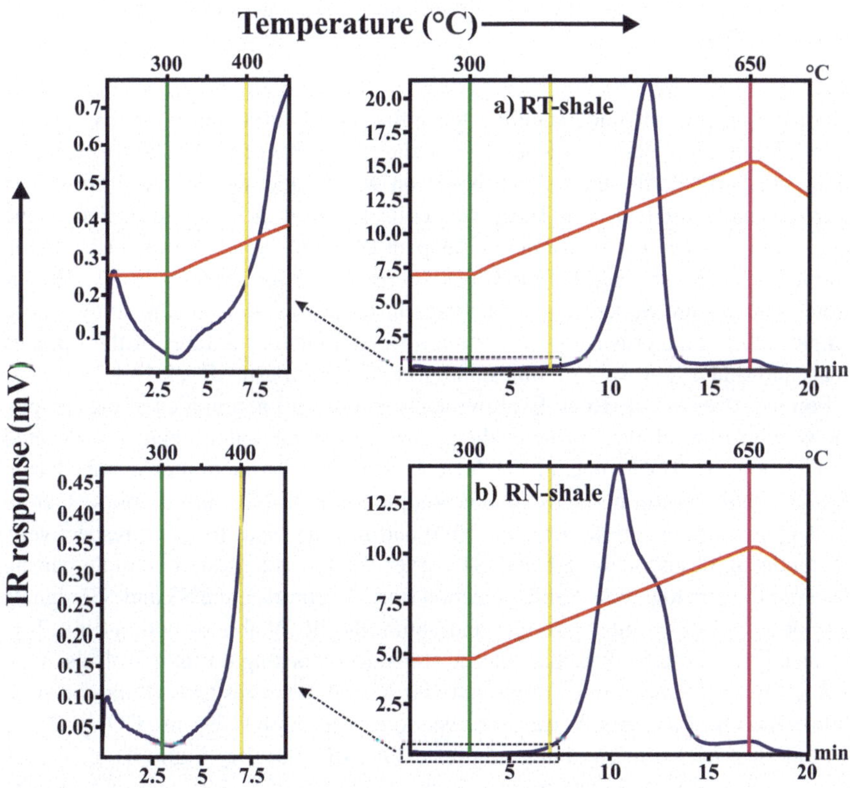

Fig. 2.7 S3 and S3' curves of two shale samples—**a** RT-shale (from Ramkola-Tatapani area of India), **b** RN-shale (from Raniganj basin, India). Note that CO_2 is released from organic-matter below 400 °C and is represented as S3. CO_2 released from siderite at temperatures > 400 °C is represented as S3'

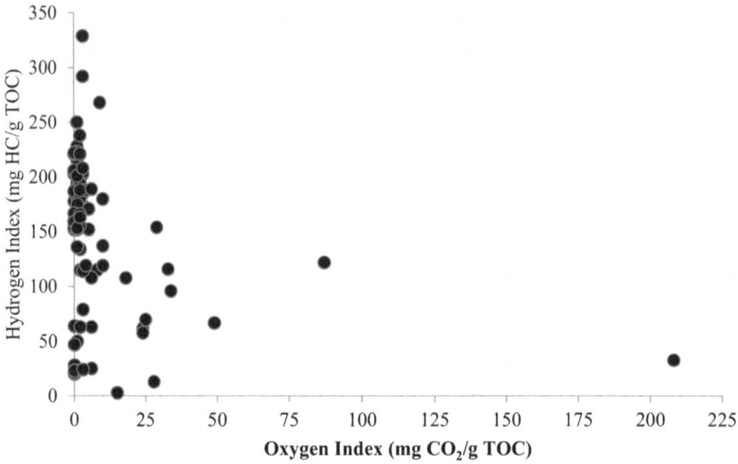

Fig. 2.8 Variation in HI and OI of Upper Carboniferous coals and shales, Poland (data from Kotarba et al. 2002)

For the purpose of understanding the influence of carbonate minerals, specifically siderite and calcite, on Rock-Eval indices, Hazra et al. (2022a, b) manually mixed crushed siderite and calcite, based on weight and simple mass ratios. The siderite sample used in their study was collected from a borehole drilled in the coal-bearing succession of Ramkola-Tatapani coalfield, South Rewa Basin, India. Similarly, a limestone sample from Chandi Formation (Mesoproterozoic age, Raipur Group, Chattisgarh Supergroup), Chattisgarh, India was used for the study. These samples after being crushed to < 212 microns, and binary mixtures of the samples (coals and shales) and siderite/limestone, were analysed using Rock-Eval.

Figure 2.9 shows the Rock-Eval pyrolysis and oxidation graphics of the sample. The siderite-rich sample, owing to the presence of some organic-matter associated with it, showed a TOC content of 2.15 wt %. Although the maximum signals during pyrolysis and oxidation stages of this sample were within the reliable detection limits of Flame Ionization Detector (FID) and Infrared detector (IR), respectively, substantial portions of the S2 and S4 curves for the siderite-rich sample can be observed to be below the reliable detection limit. In contrast, the S3 and S3' signals were observed to be much stronger and within the IR reliable detection limit. It is reasonably well established that siderite starts decomposing during pyrolysis-stage of Rock-Eval at temperatures of around 400 °C and generates CO_2 (Lafargue et al. 1998). The OI of the sample was observed to be very high (1692 mg CO_2/g TOC). The siderite-rich sample (having minimal TOC, RC, but very high OI), generates some amount of CO_2 below temperatures of 400 °C (Fig. 2.9), and indicates its influence on S3, and to the calculation of OI.

As mentioned previously, the siderite rich sample was mixed with coals. Figure 2.10 shows the S3 and S3' curves of vitrains (lithotypes in coal) separated

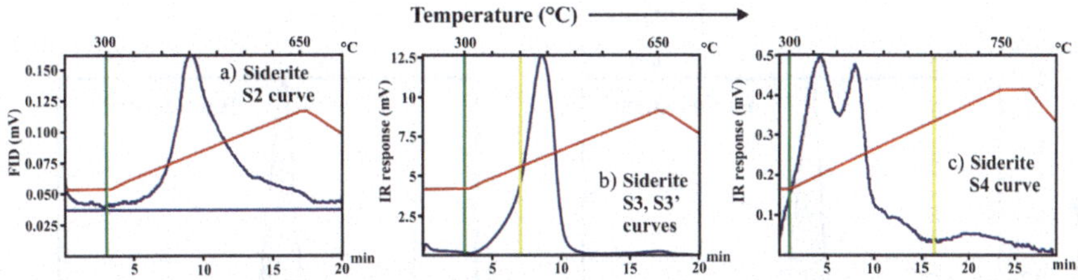

Fig. 2.9 S2, S3CO_2, S3'CO_2 and S4CO_2 curves of the siderite-rich sample and limestone sample. Note that CO_2 below 400 °C is considered as S3, while CO_2 released at temperatures > 400 °C is represented as S3'

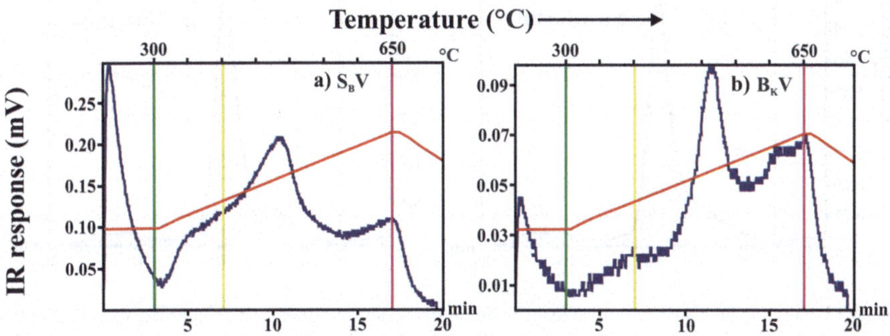

Fig. 2.10 S3 and S3' curves of the manually isolated vitrains S_BV (**a**) and B_KV (**b**). Note that CO_2 below 400 °C is considered as S3, while CO_2 released at temperatures > 400 °C is represented as S3'

manually from two coals (with distinct ranks) from India. The sample S_BV represents vitrain isolated from a high volatile bituminous coal from Sonepur Bazari area of Raniganj basin, India, while sample B_KV represents vitrain isolated from a low volatile coal from B&K area of East Bokaro basin, India. Coals from these basins were earlier studied by Hazra et al. (2021a, b). It can be seen that for sample S_BV, the S3 signal is above the minimum detection limit of Rock-Eval IR i.e., the signal is reliable. As sample S_BV is low rank (HvbA coal), greater amounts of oxygenated compounds are present within them, which are cracked during pyrolysis stage and represented under S3 curve of Rock-Eval. In contrast, for the Lvb coal B_KV, the S3 signal can be observed to lower than the reliable detection limit of Rock-Eval (Fig. 2.10). The IRCO_2 lower detection limit of Rock-Eval 6 apparatus is 0.2 mV. For all the samples, the S3' curve (indicative of presence of siderite), can be observed to be < 0.2 mV, pointing towards the absence of siderite in the studies samples. S3' for S_BV and B_KV were found to be 15.2 and 6.2 mg CO_2/g rock, while the OI of the S_BV and B_KV samples were found to be 12 and 1 mg CO_2/g TOC, respectively.

For both the samples with addition of the siderite-rich sample in simple mass ratios, a rise in S3' and S3 was observed. For example, for the S_BV sample, while the initial S3 and S3' were 8.01 and 15.2 mg CO_2/g rock, respectively, when 20%

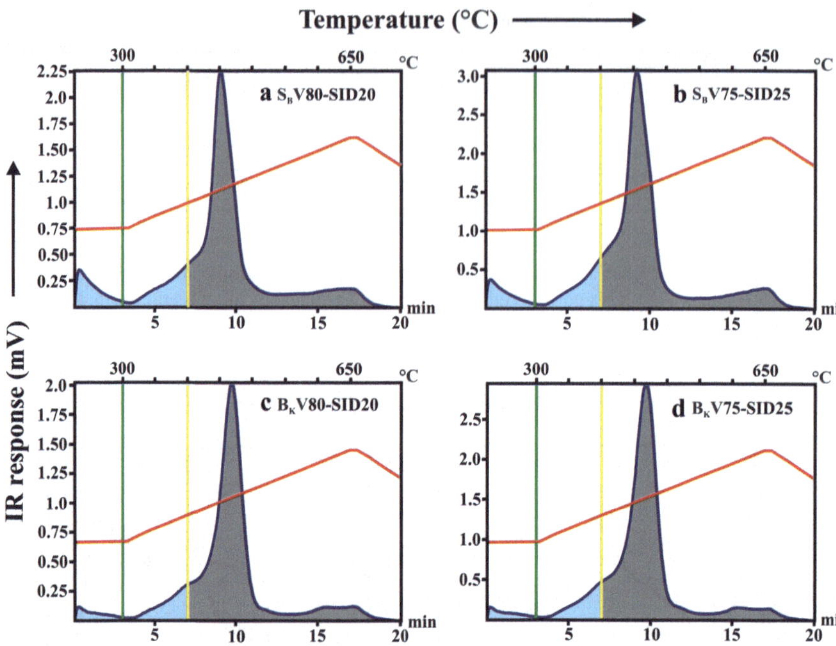

Fig. 2.11 S3CO_2 and S3'CO_2 curves of the manually isolated vitrains (S_BV and B_KV) and siderite mixtures. **a** S_BV—80 wt % and siderite—20 wt %; **b** S_BV—75 wt % and siderite—25 wt %; **c** B_KV—80 wt % and siderite—20 wt %; **d** B_KV—75 wt % and siderite—25 wt %

of siderite-rich sample was mixed with the sample, the above-mentioned parameters were increased to 12.04 and 48.2 mg CO_2/g rock, respectively (Hazra et al. 2022a, b). Consequently, the OI increased to 21 mg CO_2/g TOC. Similarly, when 25% of siderite-rich sample was mixed S_BV sample the S3, S3' and OI were observed to increase to 13.79 mg CO_2/g rock, 61.6 mg CO_2/g rock and 27 mg CO_2/g TOC, respectively. Similar increase in S3, S3' and OI with addition of siderite was noted for the B_KV sample as well. While the increase in S3' is understandable, as siderite decomposes at temperatures of around 400 °C and generates CO_2 (Lafargue et al. 1998), the increase in S3 is noteworthy. The increase in S3 with addition of siderite essentially implies that siderite decomposes at temperatures below 400 °C, and thus affects the CO_2 generated from organic matter pyrolysis. This leads to an overestimation of both S3 and OI. Unlike the low signals of both S3 and S3' noted in the raw samples, with the addition of the siderite-rich sample, the signals of both were observed to increase with values being much higher than the lower detection limits of IR detector (Fig. 2.11). What this essentially means is that due to presence of siderite the Rock-Eval OI may get severely affected and consequently could potentially result in misleading interpretations. Higher OI values generally indicate that the sample is oxidized or lower rank organic matter is present within them. As correctly inferred by Kotarba et al. (2002), an unusually high OI is associated with pyrolysis stage decomposition of certain mineral matter. Thus, OI from Rock-Eval

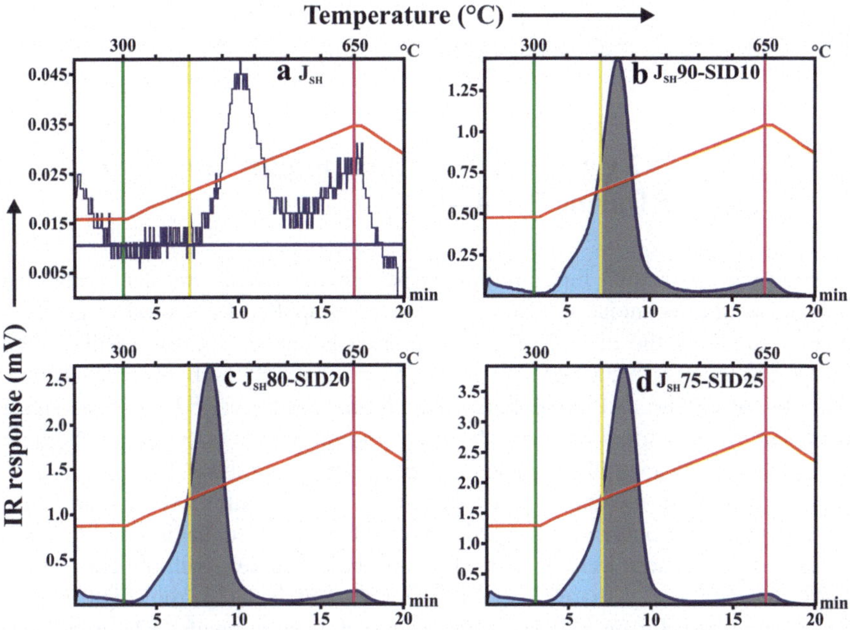

Fig. 2.12 $S3CO_2$ and $S3'CO_2$ curves of over-mature shale sample (J_{SH}) and siderite mixtures. **a** J_{SH}—100 wt %; **b** J_{SH}—90 wt % and siderite—10 wt %; **c** J_{SH}—80 wt % and siderite—20 wt %; **d** J_{SH}—75 wt % and siderite—25 wt %

may get significantly affected by the presence of siderite within the samples. What this essentially implies is that Rock-Eval S3 and S3' graphics must be meticulously cross-validated by analysts prior to reporting of OI data, as it may give clues towards role of siderite on OI of shales and coals.

Hazra et al. (2022a, b) also noted substantial increase in OI for shales mixed with siderite, indicating the false interpretations that siderite may induce while dealing with geochemical screening of shale plays. Figure 2.12 shows the $S3CO_2$ and $S3'CO_2$ plots of an over-mature shale sample (J_{SH}; T_{max}: 511 °C) from Jharia basin is with low HI (38 mg HC/g TOC) and TOC content (2.78 wt %). For the shale, while the initial S3 and S3' of the raw sample were 0.15 and 1.30 mg CO_2/g rock (Fig. 2.12a), after adding only 10 wt % of siderite to the sample (Fig. 2.12b), the same were observed to increase to 5.45 and 16.30 mg CO_2/g rock. Similarly, when the percentage of siderite was increased to 20 and 25%, the S3 and S3' were observed to increase to 8.38 and 28.00 mg CO_2/g rock, and 10.62 and 38.20 mg CO_2/g rock, respectively (Hazra et al. 2022a, b). The observed increase in both S3 and S3', for both shale and coal samples, provides clear evidence on siderite decomposing at temperatures below 400 °C during pyrolysis. This interference with the CO_2 signal from organic matter leads to an overestimation of both S3 and OI.

2.5 Nature of S2 Curves of Coals and Insights into Their Behaviour During Pyrolysis

Beyond their application for conventional petroleum systems and unconventional shale plays, the technique of Rock-Eval has been applied by several researchers for understanding the oil and/or gas generation potential of coals (Wilkins and George 2002; Petersen 2006). Very recently, Hazra et al. (2021a, b) observed distinct responses of coking and non-coking coals during Rock-Eval S2 pyrolysis stage, which led them to stress on carefully inspecting the nature of S2 curves because they provide hints about the behaviour of coals during pyrolysis. For vitrain lithotypes isolated from medium volatile (Mvb) and low volatile (Lvb) Indian coals, they observed development of spiky S2 curves. They observed that at several places, spikes and unevenness were developed within the S2 curves, with the crest of the crests being unusually sharp. In contrast, for vitrains isolated from high volatile (HvbA) coals and for the IFP Rock-Eval standard, smooth regular S2 curves were observed.

Figure 2.13a, b shows the S2 curves of vitrains isolated from two non-coking HvbA Indian coals from Sonepur Bazari (Raniganj Formation, Raniganj Basin) and Pakri Barwadih (Barakar Formation, North Karanpura Basin). For both these samples, smooth S2 curves were noted, indicating the usual processes of pyrolyzate cracking, hydrocarbon elution and detection of them by the FID. However, for the vitrains separated from coking Mvb coals, spikes, unevenness and ruggedness in the S2 curves were observed (indicated by arrows). Figure 2.13c, d shows the S2 curves of a vitrain sample isolated from a coking Mvb coal from Tisra mines (Barakar Formation, Jharia Basin), while Fig. 2.13e, f shows the S2 curves of a vitrain sample isolated from a coking Mvb coal from Salanpur mines (Barakar Formation, Raniganj Basin). To eliminate the likelihood of low sample mass being accountable for the seen roughness in the S2 curves, the vitrains isolated manually from the coking coals were re-examined at higher sample weights (~ 10 mg). The spiky and uneven characteristic of the S2 curves for these samples were found to be persistent even at the higher weights (Fig. 2.13b, d). The figure clearly shows the contrasting differences in the S2 curve shape for the coals of distinct ranks, and in a way indicates the different processes operative for coking and non-coking coals during pyrolysis stage.

To determine the causes behind formation of smooth S2 curves for non-coking coals and spiky S2 curves for coking coals, Hazra et al. (2021a) re-analysed the samples in Rock-Eval using only pyrolysis cycles and collected the residues generated after pyrolysis from the crucibles. These residues were carefully collected,

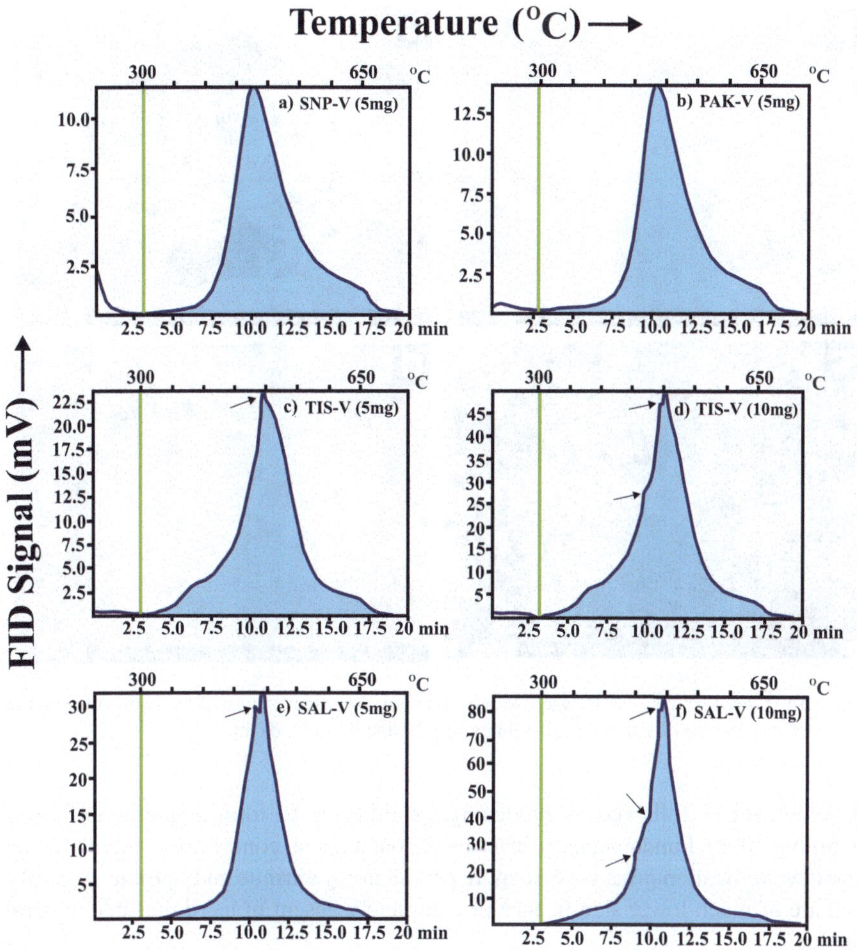

Fig. 2.13 S2 pyrograms of vitrains isolated from HvbA coals (**a**, **b**), and Mvb coals (**c**–**f**). Note the arrows indicating the development of spikes in S2 curves of Mvb vitrains. SNP-V, PAK-V, TIS-V and SAL-V denotes vitrains isolated from Sonepur Bazari, Pakri Barwadih, Tisra and Salanpur areas of India

pasted in carbon tape, subsequently gold coated and then examined under FE-SEM. Under the FE-SEM contrasting morphologies were observed for the isolated vitrains from the Mvb-Lvb coals and HvbA coals.

When the pyrolysis products of the coking Mvb coal vitrains from Tisra and Salanpur mines (S2 curves of which are shown in Fig. 2.13c–d and e–f) were observed under FE-SEM, solid-lump residues with clear presence of flow textures and vesicles were observed (Fig. 2.14). These textures typically result when coking coals are heated under inert conditions. When heating without the presence of oxygen, coking coals first soften, then melt and mimics a plastic nature, with concomitant

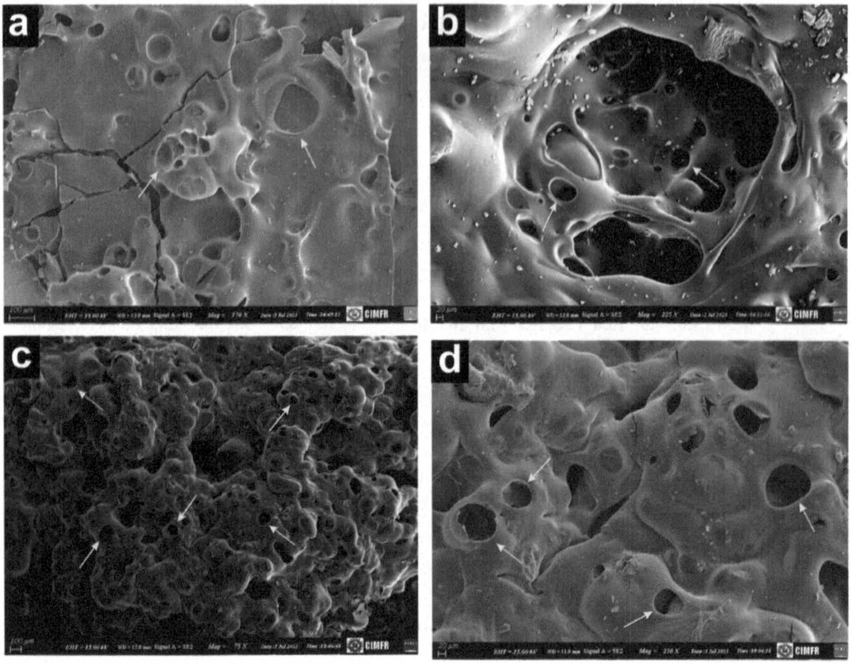

Fig. 2.14 Flow textures and 3D vacuoles (from bubble bursting; indicated by arrows) formed in the pyrolysis residue of vitrains from Salanpur (**a**, **b**) and Tisra (**c**, **d**) mines

devolatilization, followed by hardening/resolidifying to form a sponge-like mass (Crelling 2008). Fundamentally, in terms of coal maceral composition, these melting and devolatilization processes are more prevalent for vitrinite and liptinite macerals, and are of much lower significance or completely absent in inertinites and mineral matter.

The development of the three-dimensional irregular vacuoles, as indicated by arrows in Fig. 2.14, were explained by Hazra et al. (2021a) to be formed due to bursting of gaseous bubbles caused due to melt generation and irregular release of gases during pyrolysis, prior to final hardening or re-solidification. Consequently, the spikes formed in the S2 curves are also attributed to the volatile gases occasionally or sporadically released from the molten mass when gaseous bubbles burst. In contrast to the coking Mvb vitrains, for the non-coking HvbA vitrains from Sonepur Bazari and Pari Barwadih mines (S2 curves of which are shown in Fig. 2.13a, b), a pulverulent residue was noted with well-developed/sharp grain boundaries, showing no signs of melting or development of flow textures and bubble bursting induced vacuoles (Fig. 2.15). Consequently, for the non-coking coals as no bubble bursting happened, the S2 curves produced from Rock-Eval pyrolysis stage are smooth and regular. Importantly, the development of uneven/spiky S2 for coking vitrains, and the lack of the same for non-coking-vitrains, puts forward the importance of S2 curve shapes as

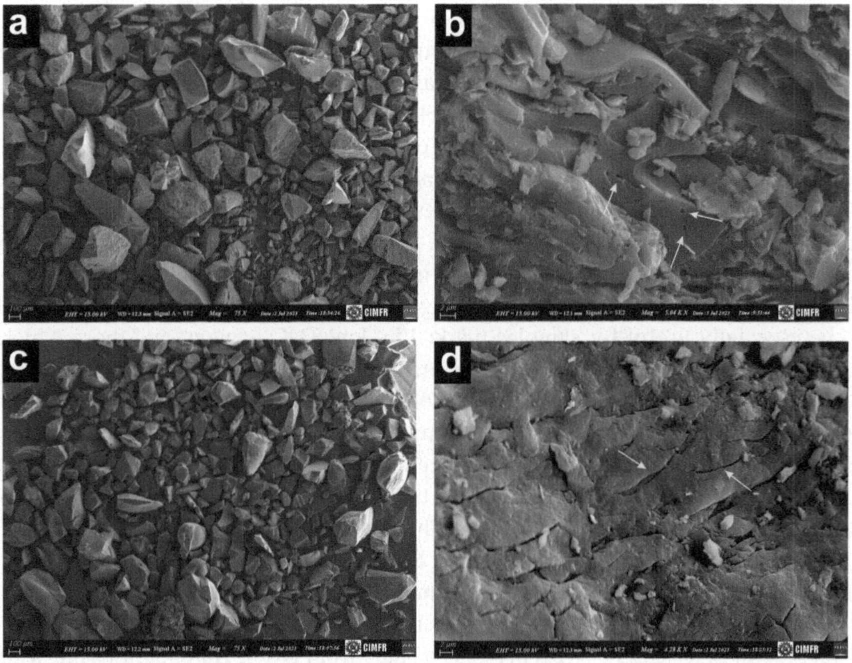

Fig. 2.15 Pulverulent residue formed after pyrolysis of vitrains from Sonepur Bazari (**a**, **b**) and Pakri Barwadih (**c**, **d**) mines. The arrows indicate formation of pores/cracks in the samples, which are distinctly different from those showed in Fig. 2.14

it can provide detailed and significant insights into the behaviour of coals, and can help in predicting better end-utilization of coals.

2.6 Using Rock-Eval S4T_{peak} for Estimating Thermal Maturity

T_{max}, calculated from Rock-Eval S2 curve maxima, is widely used as a proxy for estimating thermal maturity levels of sedimentary organic matter. However, under certain conditions, T_{max} may yield unreliable results. While Peters (1986) observed lowering of T_{max} in a marginally mature carbonaceous shale with high resinite concentration, Snowdon (1995) observed T_{max} lowering in certain rocks with high HI values. On the other hand, Copard et al. (2002) for a group of weathered coal samples substantiated erroneous maturity assessment given by Rock-Eval T_{max}. Very recently, Carjaval-Ortiz and Gentzis (2015) documented that for type I kerogen rich samples, i.e., for samples with high pyrolyzate yield, at higher sample weights as greater amounts of hydrocarbons are generated beyond the detection limit of the instrument (125 kV), the S2 curve becomes broader and T_{max} may be wrongly calculated. Similarly, for samples with minimal organic matter or over mature organic matter, very low hydrocarbons may be generated, below the instrument detection limit (0.1 mV) which can also result in erroneous thermal maturity assessment (Hazra et al. 2021a).

To overcome these challenges, Hazra et al. (2021a) recently used the Rock-Eval $S4T_{peak}$ as a proxy for thermal maturity of shales. For a set of samples spanning across different thermal maturity levels and organic composition from India, the temperature peak of the $S4CO_2$ curve generated from combustion of the organic matter during Rock-Eval oxidation stage, was noted to progressively rise with the advancement of thermal maturity levels, indicating its usefulness is predicting the thermal maturity levels of sedimentary organic matter. They observed the S4Tpeak to be especially useful for samples marked by low TOC content, and those which underwent thermal metamorphism. For such samples, they observed the S2 curves to be unreliable, showing uneven morphology and having low FID signals. Even further, Hazra et al. (2021a) reported strong correlation of the $S4T_{peak}$ values to the temperature maxima obtained from Differential Scanning Calorimetry experiments, and in their study calculated $S4T_{peak}$ of both raw shales and pyrolyzed shales to establish the significance of $S4T_{peak}$ as a thermal maturity proxy.

Figure 2.16 shows S2 pyrograms and $S4CO_2$ oxidation graphics of few shales from India. RJ denotes a shale from Rajmahal Basin of India (Fig. 2.16A1; T_{max}: 424 °C), and samples from this basin are generally marked by lower thermal maturity levels (Sethi et al. 2023). NK represents a shale from North Karanpura basin, with a T_{max} of 441 °C (early mature; Fig. 2.16B1). JH represents a shale from Jharia basin India (Fig. 2.16C1; T_{max}: 452 °C; late mature) and samples from this basin are generally marked by higher thermal maturity levels (Sethi et al. 2023). For these three shales, the S2 pyrograms can be observed to be symmetric, indicating the T_{max} to be reliable. Figure 2.16D1 shows the S2 pyrogram of a heat-altered shale from Jharia basin. For this sample, the S2 pyrogram can be observed to show a wider disposition, indicating unreliable T_{max} (T_{max}: 561 °C). Moreover, for this sample, the FID signal can be observed to be low, compared to the other shales. This phenomenon is primarily induced by the release of hydrocarbons from the sample, resulting from the influence of the igneous intrusion. Fig. 2.16A2, B2, C2 and D2 shows the $S4CO_2$ oxidation graphics of the RJ, NK, JH, and heat-altered shales. It is evident that as the thermal maturity levels of the samples increase, the $S4T_{peak}$ systematically increases, indicating its suitability in being used as a proxy for thermal maturity. It is important to realize that while the S4Tpeak shows a step-by-step increase with increasing thermal maturity of the samples (Fig. 2.16), it should only be used in place of T_{max} (from S2) when the S2 curves are unreliable, asymmetric, broad and the FID signals are too low.

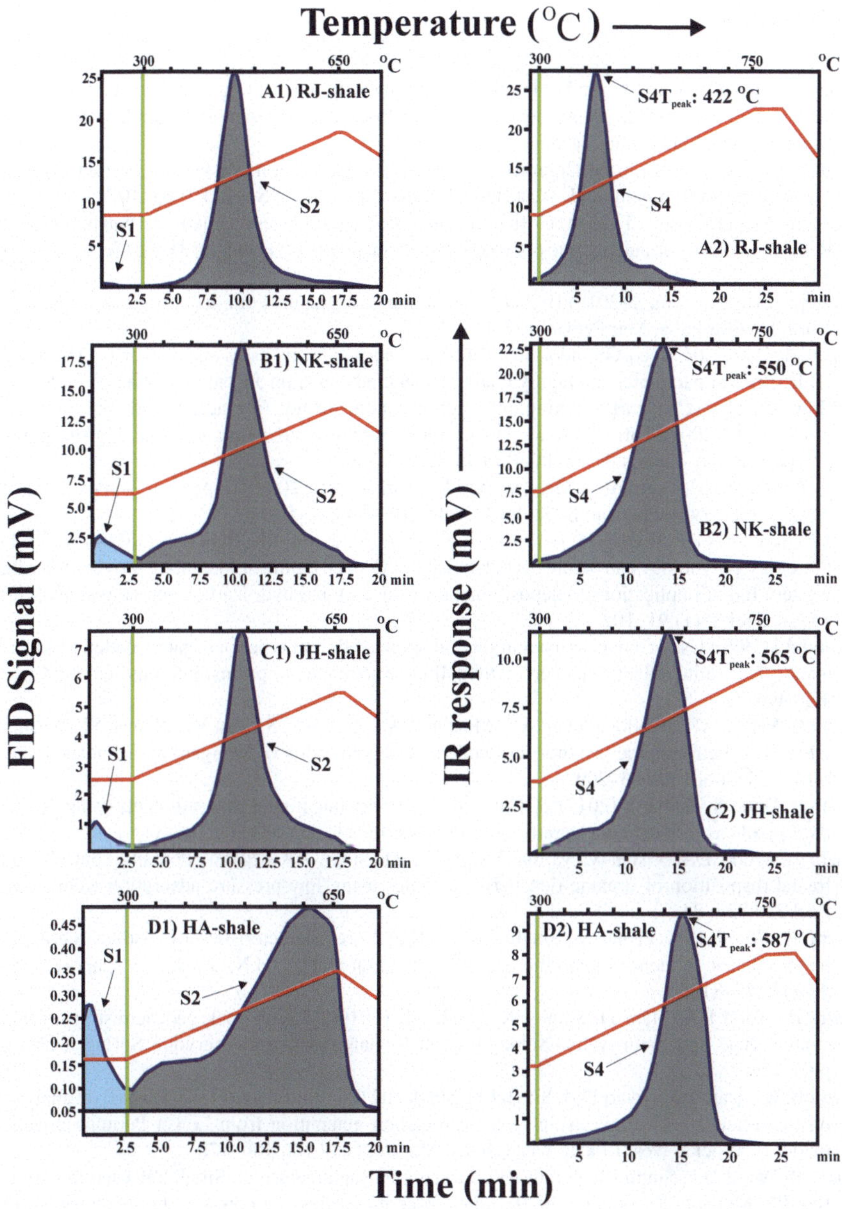

Fig. 2.16 S2 pyrograms (**A1**, **B1**, **C1**, **D1**) and S4CO_2 oxidation graphics (**A2**, **B2**, **C2**, **D2**) of shales with contrasting thermal maturity levels

References

Behar F, Vandenbroucke M (1987) Chemical modelling of kerogens. Org Geochem 11:15–24

Behar F, Beaumont VDEB, Penteado HDB (2001) Rock-Eval 6 technology: performances and developments. Oil Gas Sci Technol 56(2):111–134

Boruah A, Ganapathi S (2015) Organic richness and gas generation potential of Permian Barren Measures from Raniganj field, West Bengal, India. J Earth Syst Sci 124:1063–1074

Carvajal-Ortiz H, Gentzis T (2015) Critical considerations when assessing hydrocarbon plays using Rock-Eval pyrolysis and organic petrology data: data quality revisited. Int J Coal Geol 152:113–122

Chattopadhyay A, Dutta S (2014) Higher plant biomarker signatures of Early Eocene sediments of North Eastern India. Mar Pet Geol 57:51–67

Copard Y, Disnar JR, Becq-Giraudon JF (2002) Erroneous maturity assessment given by Tmax and HI Rock-Eval parameters on highly mature weathered coals. Int J Coal Geol 49(1):57–65

Crelling JC (2008) Coal carbonization. In: Applied coal petrology. Elsevier, pp 173–192

Dembicki H Jr (2009) Three common source rock evaluation errors made by geologists during prospect or play appraisals. AAPG Bull 93(3):341–356

Dutta S, Mallick M, Bertram N, Greenwood PF, Mathews RP (2009) Terpenoid composition and class of Tertiary resins from India. Int J Coal Geol 80(1):44–50

Dutta S, Mathews RP, Singh BD, Tripathi SM, Singh A, Saraswati PK, Banerjee S, Mann U (2011) Petrology, palynology and organic geochemistry of Eocene lignite of Matanomadh, Kutch Basin, western India: implications to depositional environment and hydrocarbon source potential. Int J Coal Geol 85(1):91–102

Espitalié J (1986) Use of Tmax as a maturation index for different types of organic matter. Comparison with vitrinite reflectance, thermal modelling in sedimentary basins. Editions Technip Paris, 475–496

Hazra B, Varma AK, Bandopadhyay AK, Mendhe VA, Singh BD, Saxena VK, Samad SK, Mishra DK (2015) Petrographic insights of organic matter conversion of Raniganj basin shales, India. Int J Coal Geol 150:193–209

Hazra B, Dutta S, Kumar S (2017) TOC calculation of organic matter rich sediments using Rock-Eval pyrolysis: critical consideration and insights. Int J Coal Geol 169:106–115

Hazra B, Wood DA, Vishal V, Varma AK, Sakha D, Singh AK (2018a) Porosity controls and fractal disposition of organic-rich Permian shales using low-pressure adsorption techniques. Fuel 220:837–848

Hazra B, Wood DA, Vishal V, Singh AK (2018b) Pore characteristics of distinct thermally mature shales: influence of particle size on low-pressure CO_2 and N_2 adsorption. Energy Fuels 32(8):8175–8186

Hazra B, Wood DA, Mani D, Singh PK, Singh AK (2019a) Source-rock geochemistry: organic content, type, and maturity. In: Evaluation of shale source rocks and reservoirs. Springer, Cham, pp 7–17

Hazra B, Karacan CÖ, Tiwari DM, Singh PK, Singh AK (2019b) Insights from Rock-Eval analysis on the influence of sample weight on hydrocarbon generation from Lower Permian organic matter rich rocks, West Bokaro basin, India. Mar Pet Geol 106:160–170

Hazra B, Wood DA, Singh PK, Singh AK, Kumar OP, Raghuvanshi G, Singh DP, Chakraborty P, Rao PS, Mahanta K, Sahu G (2020) Source rock properties and pore structural framework of the gas-prone Lower Permian shales in the Jharia basin, India. Arab J Geosci 13:1–18

Hazra B, Singh DP, Crosdale PJ, Singh V, Singh PK, Gangopadhyay M, Chakraborty P (2021a) Critical insights from Rock-Eval analysis of vitrains. Int J Coal Geol 238:103717

Hazra B, Singh DP, Chakraborty P, Singh PK, Sahu SG, Adak AK (2021b) Using rock-eval S4Tpeak as thermal maturity proxy for shales. Mar Pet Geol 127:104977

Hazra B, Singh DP, Chakraborty P, Das H, Singh V, Sahu SG, Singh PK (2022a) Structural and thermal properties of vitrain lithotype in coal-inferences from TG-DTG-DSC, Rock-Eval and X-ray diffraction. J Earth Syst Sci 131(2):98

Hazra B, Katz BJ, Singh DP, Singh PK (2022b) Impact of siderite on Rock-Eval S3 and oxygen index. Mar Pet Geol 143:105804

Inan S, Yalçin MN, Mann U (1998) Expulsion of oil from petroleum source rocks: inferences from pyrolysis of samples of unconventional grain size. Org Geochem 29(1–3):45–61

İnan S, Henderson S, Qathami S (2017) Oxidation Tmax: a new thermal maturity indicator for hydrocarbon source rocks. Org Geochem 113:254–261

Jarvie DM (1991) Total organic carbon (TOC) analysis. In: Merrill RK (ed) Source and migration processes and evaluation techniques. American Association of Petroleum Geologists, Treatise of Petroleum Geology, pp 113–118

Jarvie DM (2012a) Shale resource systems for oil and gas: Part 1—Shale–gas resource systems; In: Breyer JA (ed) Shale reservoirs—giant resources for the 21st century. American Association of Petroleum Geologists Memoir, pp 97 69–87

Jarvie DM (2012b) Shale resource systems for oil and gas: Part 2—Shale–gas resource systems; In: Breyer JA (ed) Shale reservoirs—giant resources for the 21st century. American Association of Petroleum Geologists Memoir, pp 97 89–119

Katz BJ (1983) Limitations of 'Rock-Eval' pyrolysis for typing organic matter. Org Geochem 4(3–4):195–199

Kotarba MJ, Clayton JL, Rice DD, Wagner M (2002) Assessment of hydrocarbon source rock potential of Polish bituminous coals and carbonaceous shales. Chem Geol 184(1–2):11–35

Lafargue E, Marquis F, Pillot D (1998) Rock-Eval 6 applications in hydrocarbon exploration, production, and soil contamination studies. Revue De L'institut Français Du Pétrole 53(4):421–437

Li M, Chen Z, Qian M, Ma X, Jiang Q, Li Z, Tao G, Wu S (2020) What are in pyrolysis S1 peak and what are missed? petroleum compositional characteristics revealed from programed pyrolysis and implications for shale oil mobility and resource potential. Int J Coal Geol 217:103321

Mani D, Patil DJ, Dayal AM (2015) Organic properties and hydrocarbon generation potential of shales from few sedimentary basins of India. Petroleum Geosci: Indian Contexts, 99–126

Mathews RP, Singh BD, Singh VP, Singh A, Singh H, Shivanna M, Dutta S, Mendhe VA, Chetia R (2020) Organo-petrographic and geochemical characteristics of Gurha lignite deposits, Rajasthan, India: insights into the palaeovegetation, palaeoenvironment and hydrocarbon source rock potential. Geosci Front 11(3):965–988

Mathews RP, Singh BD, Singh VP (2018a) Evaluation of organic matter, hydrocarbon source, and depositional environment of onshore Warkalli sedimentary sequence from Kerala-Konkan Basin, south India. J Geol Soc India 92(4):407–418

Mathews RP, Singh BD, Singh H, Singh VP, Singh A (2018b) Characterization of Panandhro Lignite deposits (Kachchh Basin), western India: results from the bulk geochemical and palynofloral compositions. J Geol Soc India 91(3):281–289

Paul S, Sharma J, Singh BD, Saraswati PK, Dutta S (2015) Early Eocene equatorial vegetation and depositional environment: biomarker and palynological evidences from a lignite-bearing sequence of Cambay Basin, western India. Int J Coal Geol 149:77–92

Peters KE (1986) Guidelines for evaluating petroleum source rock using programmed pyrolysis. AAPG Bull 70(3):318–329

Peters KE, Cassa MR (1994) Applied source rock geochemistry. In: Magoon LB, Dow WG (eds) The petroleum system-from source to trap. American Association of Petroleum Geologists, pp 93–120

Petersen HI (2006) The petroleum generation potential and effective oil window of humic coals related to coal composition and age. Int J Coal Geol 67(4):221–248

Romero-Sarmiento MF, Pillot D, Letort G, Lamoureux-Var V, Beaumont V, Huc AY, Garcia B (2016) New Rock-Eval method for characterization of unconventional shale resource systems. Oil & Gas Sci Tech-Revue d'IFP Energies Nouvelles 71(3):37

Sethi C, Hazra B, Wood DA, Singh AK (2023) Experimental protocols to determine reliable organic geochemistry and geomechanical screening criteria for shales. J Earth Syst Sci 132(2):45

Singh AK, Kumar A (2020) Assessment of thermal maturity, source rock potential and paleodepositional environment of the paleogene lignites in Barsingsar, Bikaner-Nagaur Basin, Western Rajasthan, India. Nat Resour Res 29:1283–1305

Singh VP, Singh BD, Singh A, Singh MP, Mathews RP, Dutta S, Mendhe VA, Mahesh S, Mishra S (2017a) Depositional palaeoenvironment and economic potential of Khadsaliya lignite deposits (Saurashtra Basin), western India: based on petrographic, palynofacies and geochemical characteristics. Int J Coal Geol 171:223–242

Singh VP, Singh BD, Mathews RP, Singh A, Mendhe VA, Singh PK, Mishra S, Dutta S, Shivanna M, Singh MP (2017b) Investigation on the lignite deposits of Surkha mine (Saurashtra Basin, Gujarat), western India: their depositional history and hydrocarbon generation potential. Int J Coal Geol 183:78–99

Singh AK, Hakimi MH, Kumar A, Ahmed A, Abidin NSZ, Kinawy M, Mahdy OE, Lashin A (2020) Geochemical and organic petrographic characteristics of high bituminous shales from Gurha mine in Rajasthan, NW India. Scient Rep 10(1):22108

Singh DP, Singh V, Singh PK, Hazra B (2021) Source rock properties and pore structural features of distinct thermally mature Permian shales from South Rewa and Jharia basins, India. Arabian J Geosci 14(10):916

Singh DP, Wood DA, Singh V, Hazra B, Singh PK (2022) Impact of particle crush-size and weight on Rock-Eval S2, S4, and kinetics of shales. J Earth Sci 33(2):513–524

Snowdon LR (1995) Rock-Eval Tmax suppression: documentation and amelioration. AAPG Bull 79(9):1337–1348

Sykes R, Snowdon LR (2002) Guidelines for assessing the petroleum potential of coaly source rocks using Rock-Eval pyrolysis. Org Geochem 33(12):1441–1455

Taylor GH, Davis AT, Diessel CFK, Littke R, Robert P (1998) Organic petrology. In: Gerbrüder Borntraeger. Berlin, p 704

Tissot BP, Welte DH (1978) Petroleum formation and occurrence; a new approach to oil and gas exploration. Springer-Verlag, Berlin, Heidelberg, New York

Varma AK, Hazra B, Mendhe VA, Chinara I, Dayal AM (2015a) Assessment of organic richness and hydrocarbon generation potential of Raniganj basin shales, West Bengal, India. Mar Pet Geol 59:480–490

Varma AK, Biswal S, Hazra B, Mendhe VA, Misra S, Samad SK, Singh BD, Dayal AM, Mani D (2015b) Petrographic characteristics and methane sorption dynamics of coal and shaly-coal samples from Ib Valley Basin, Odisha, India. Int J Coal Geol 141:51–62

Varma AK, Khatun M, Mendhe VA, Hazra B, Singh BD, Dayal AM (2015c) Petrographic characterization and Langmuir volume of shales from Raniganj coal basin, India. J Geol Soc India 86:283–294

Wilkins RW, George SC (2002) Coal as a source rock for oil: a review. Int J Coal Geol 50(1–4):317–361

Wood DA, Hazra B (2018) Pyrolysis S2-peak characteristics of Raniganj shales (India) reflect complex combinations of kerogen kinetics and other processes related to different levels of thermal maturity. Adv Geo-Energy Res 2(4):343–368

Chapter 3
Estimation of Gas Storage Capacity Estimate in Coals and Shales

In the last decade, there has been a lot of research on the exploration and development of coalbed methane (CBM) and shale gas (SG), two sources of substantial unconventional natural gas resources (Moore 2012; Yuan et al. 2015; Zhou et al. 2016; Li et al. 2018; Hazra et al. 2019a, b, 2020a, b, 2022; Vishal et al. 2019; Chandra et al. 2023). A wealth of data has demonstrated that the CBM and SG in reservoirs are highly reliant on a variety of factors, such as the physical characteristics of the reservoir, the composition of the material, temperature and pressure conditions, burial depth, hydrodynamic conditions, depositional environment, and geologic structures (Moore 2012; Yuan et al. 2015; Zhou et al. 2016; Li et al. 2018). These factors typically cause the gas content in a reservoir to be highly lateral and vertically anisotropic. One of the main reservoir factors influencing the natural gas resource development's economic potential is gas storage. It is widely acknowledged that in in-situ reservoirs, both CBM and SG exist in adsorbed, free, and dissolved states (Pan and Connell 2015). Typically, coal reservoirs, especially those with middle- and high-rank coals, predominantly contain adsorbed gas (Liu et al. 2013). However, in low-rank coals, it is crucial not to overlook both free and dissolved gas (Liu et al. 2013). In shale reservoirs, unlike coalbed methane (CBM), both the adsorbed and free states of shale gas (SG) coexist, with adsorbed gas constituting 20–85% of the total gas (Hill and Nelson 2000). In shales with high thermal maturity, the contribution of dissolved gas can be disregarded (Li et al. 2018). In middle- and high-rank coals as well as shales, adsorbed and free gas collectively dominate, with a notably higher concentration of free gas in shales compared to coals. The contents of both adsorbed and free gas directly influence gas production due to distinct transport mechanisms (Wu et al. 2015, 2016), emphasizing the importance of determining total, adsorbed, and free gas contents for the production of shale gas (SG) and coalbed methane (CBM) (Pan and Connell 2015). The determination of total gas content involves estimating various gas components, including adsorbed, free, and dissolved gas. The Langmuir isothermal adsorption equation is commonly employed to estimate adsorbed gas content (Yi et al. 2019). This equation delineates the quantitative correlation

B. Hazra et al., *Unconventional Hydrocarbon Reservoirs: Coal and Shale*, Petroleum Engineering, https://doi.org/10.1007/978-3-031-53484-3_3

between adsorbed gas and changing pressures at a consistent temperature, offering the theoretical maximum gas adsorption or storage capacity for coal or shale. By assessing these gas components, a comprehensive understanding of total gas content is achieved, essential for the evaluation and optimization of gas production in both coal and shale formations.

3.1 Mechanisms of Adsorption

Three distinct thermodynamic states of gas are present in shale formations: as free gas within fractures and intergranular pores, as adsorbed gas on organic matter and clay minerals, and as dissolved gas within bitumen and kerogen (Schettler and Parmoly 1990; Zhang et al. 2012). Adsorption is the process by which molecules of gas, referred to as the adsorbate, accumulate on the surface of organic matter within shale because of interfacial surface forces, with kerogen serving as the adsorbent. Physisorption is governed by the prevalent dispersion-repulsion forces, namely van der Waals forces. In contrast, chemisorption involves significantly stronger forces arising from electron transfer or sharing, providing a distinctive mechanism for adsorption (Ruthven 2006). Physisorption exhibits lower enthalpy, approximately 1–1.5 times the latent heat of evaporation and is generally alterable without causing structural changes in the adsorbent. Interaction forces limit chemisorption to fewer than one layer, while physisorption usually occupies micropores at low pressures and forms multilayers of condensed gas molecules on the surfaces of meso- and macropores at higher pressures (Busch and Gentserblum 2011). Physical adsorption prevails as the fundamental mechanism for gas storage in shale rich in organic matter (Kang et al. 2011). Shales are renowned for their intricate microporous network and a broad range of pore sizes (Clarkson et al. 2013; Busch and Gentserblum 2011). Three primary categories can be used to classify the pore system found in coal and shales, which is crucial to the gas adsorption process: micropores having characteristic widths below 2 nm, mesopores ranging from 2 to 50 nm, and macropores that are larger than 50 nm (Gregg et al. 1967; Perera et al. 2012). Within these classifications, micropores play a dominant role in gas sorption, as micropores are nearly linked to the coal mass's total pore volume (Gurdal et al. 2001). This characteristic imparts coal with elevated surface areas and greater gas sorption capacities contrasted to conventional reservoirs, making it a favorable environment for gas storage (Harpalani and Schraufnagel 1990). Micropores involve a pore-filling mechanism, where gas molecules are closely packed due to the size similarity between the pores and the adsorbate molecules (Dabrowski 2001). This causes a several layers or single layer molecules of gas in the condensed phase to be accommodated in a continual filling process that is powered by the strong force fields inside the small pores (Lowell et al. 2006). In contrast, mesopores facilitate capillary adsorbate condensation, allowing gas molecules to populate these intermediate-sized pores. In micropores, the sorption capacity is predominantly influenced by the pore volume. Conversely, in mesopores, adsorption is governed by a combination of gas molecules' interactions with pore walls and attractive forces among adsorbed molecules. This dual mechanism results in two sequential adsorption stages: at low pressures, mono- and multilayer adsorption takes place first, and then capillary condensation, which leads to the final gas saturation pressure (P_0). The latter marks a notable first-order transition from gas to liquid phase, highlighting the complex dynamics of mesopore adsorption. Mesopores are characterized by porosity, specific surface area and pore size size distribution, all derived from their adsorption capacity (Sing 1985, 2001, 2004). Conversely, the surface area estimated from micropore adsorption can be misleading, as gas in place (GIP) in micropores is primarily dictated by their volume. The amount of adsorbed gas per mass of adsorbent (W) relies on equilibrium pressure (P), temperature (T), and the interaction pressure (E) between the surface molecules (adsorbent) and the gas species (adsorbate) (Gregg et al. 1967).

The quantity of gas molecules adsorbed at a constant reservoir temperature per unit weight of adsorbent is expressed as follows

$$W = f(P)_E \tag{3.1}$$

where the interaction potential (E) remains constant for a specific combination of adsorbate and adsorbent. The above equation serves as a representation of the sorption isotherm, and the graphical depiction of the relationship between adsorbent weight (W) and pressure (P) is commonly termed the isothermal curve (Yang and Liu 2020). This curve serves to illustrate the upper limit of gas adsorption onto a solid material under varying pressure conditions while maintaining a consistent temperature. De Boer's classification system further categorizes physical adsorption isotherms into six distinct types, providing valuable insights into different adsorption behaviours (Sing 1985, 2001, 2004; de Boer 1958). These six types of curves are depicted in Fig. 3.1, where the relative pressure is expressed as the ratio of equilibrium pressure to saturation pressure (P/P_0). The Langmuir adsorption model is represented by the type I isotherm, which is characterised by a plateau as it gets closer to saturation ($P/P_0 \rightarrow 1$). This isotherm is frequently seen in physisorption in microporous materials, where adsorption is limited to one or a few layers. The increase capacity of adsorption mainly depends on the available volume of micropores. The S-shaped curve that characterises the type II isotherm indicates infinite sorption behaviour. When it comes to physisorption in nonporous or macroporous materials, this sigmoid pattern is observed. A characteristic of mesoporous materials, the Type IV isotherm is very similar to the Type II isotherm. The hysteresis loop observed in a isotherm of Type IV arises due to pore condensation, with the plateau indicating the saturation of pore filling. In the initial segments of Type II and Type IV isotherms, the formation of monolayers and multilayers adsorption occurs. Point B, denoting the inflection point in both isotherms, signifies the culmination of monolayer coverage. The monolayer capacity is the term used to describe the adsorbed quantity attained at this point. Multilayer adsorption also begins at this critical point. The Type III isotherm, on the other hand, has a unique profile that is convex throughout the pressure range and lacks an inflection point. The convexity arises from relatively feeble gas–solid interactions, with attractive forces between gas molecules playing a dominant role in the prevailing force field, thereby contributing to the observed convex character of the isotherm. This isotherm is characteristic of sorption within non- or microporous solids. The Type V isotherm exhibits convexity, just like the Type III isotherm, but it has an inflection point that happens at a higher relative pressure. It is commonly observed in meso- or microporous solids characterized by weak interactions between the adsorbate and adsorbent. The presence of hysteresis loops in both Type IV and Type V isotherms provides insights into the pore network connectivity, with the upper curves indicating desorption and the lower curves representing the adsorption process. A pronounced "knee" in the desorption isotherm often suggests limited pore connectivity, which may be linked to the presence of ink-bottle-type pores (de Boer 1958; Ravikovitch et al. 1995; Neimark et al. 2000; Vishnyakov and Neimark 2003; Thommes et al. 2006). Type VI isotherm is characterized by a stepped pattern and is infrequently encountered (Sing 1985). The configuration of sorption isotherms is influenced by the intensity of interactions between the adsorbate and adsorbent, as well as interactions among adsorbate molecules. These factors are additionally influenced by the specific properties of the adsorbate, adsorbent, and the geometry of the pores within the material (Yang and Liu 2020).

3.2 Sorption Models

3.2.1 Langmuir Model

The Langmuir model is widely employed in describing shale/coal gas sorption isotherms due to its simplicity. A key assumption of the Langmuir model is the presence of a homogeneous and energetically uniform surface. As the pressure rises, the model suggests that adsorbed molecules gradually stick to the surface, forming a monolayer that covers the whole surface. It is at this point that adsorption stops. Langmuir model, designed for gas/solid adsorption in a single layer, effectively describes the isotherms of microporous materials displaying a Type I profile (Fig. 3.1).

Equation (3.2) delineates the classical Langmuir isotherm model, defined by two regression parameters, V_L and P_L. When examining the Langmuir equation, a linear relationship should be evident when plotting *P/V versus P*. The two Langmuir constants from this linear regression line are calculated using its slope and intercept from

$$\frac{P}{V} = \frac{1}{V_L}P + \frac{P_L}{V_L} \tag{3.2}$$

where fitted line's slope inversely relates to monolayer sorption capacity (V_L), while the intercept signifies the two Langmuir constants' ratio. V_L quantifies shale's gas storage capacity, whereas P_L denotes the gas's surface affinity. A higher P_L value indicates a more robust gas-surface interaction.

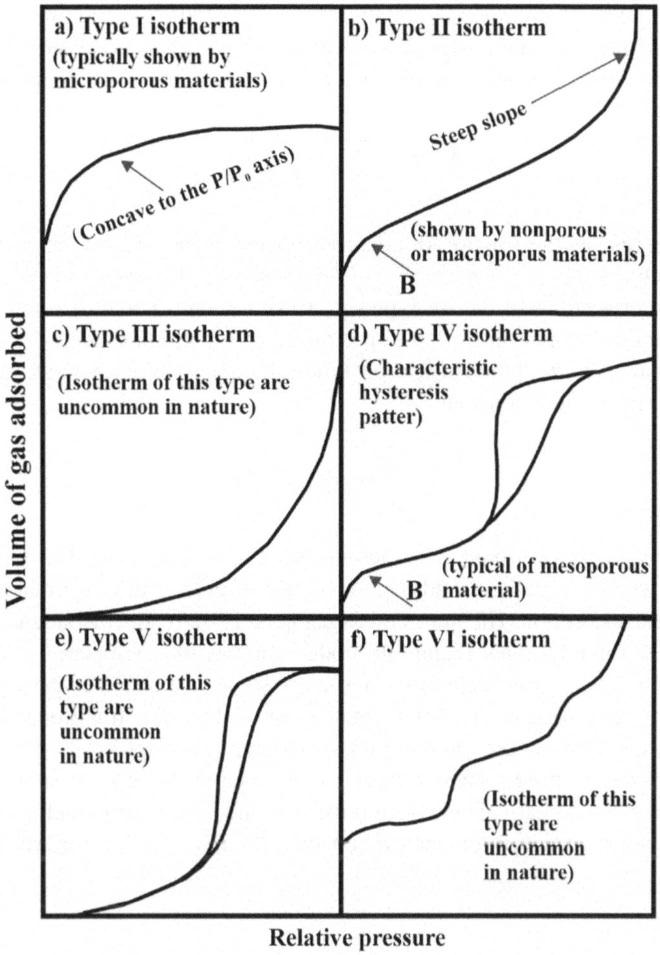

Fig. 3.1 Physical adsorption types categorised by the IUPAC (Sing 1985; Pure and applied chemistry, 57(4), pp. 603–619. Copyright © 1985 IUPAC)

The Langmuir model is a useful tool for fitting experimental sorption data despite its potential inability to fully represent the heterogeneous properties of shale due to its simplicity. For many of the microporous adsorbents (like shale), where an increase in capacity of adsorption is observed, this model successfully fits the experimental sorption isotherms. Adsorption sites in shale are classified into two categories: clay minerals or kerogen. The sorption potential of the two types of sites is distributed in two ways (bimodal distribution). The extended Langmuir model has been tailored into the bi-Langmuir model, which effectively models the adsorption isotherm of shale rich in clay minerals (Lu et al. 1995).

The extended Langmuir model (Kapoor et al. 1990) is a modification of the classical Langmuir model designed to account for the existance of an energetically heterogenous surface, and the modified equation reflects this consideration.

$$V = V_L \sum_{i=1} f_i \frac{b_i P}{1 + b_i P} \tag{3.3}$$

Here, f_i represents the fraction of adsorption sites of the ith type among the total sites characterized by V_L, where $\sum_{i=1} f_i$. b_i stands for the inverse of the Langmuir volume corresponding to the ith type of adsorption site. For $i = 2$, the extended Langmuir model transforms into the specialized bi-Langmuir model.

The physical interpretation of b, as per kinetic theory, is the temperature-dependent adsorption equilibrium constant.

$$b = \frac{b_0}{\sqrt{T}} \exp(-E/RT) \tag{3.4}$$

Here, b_0 represents a constant characteristic that remains unaffected by temperature variations, while E signifies the adsorption activation energy, often estimated through isosteric heat adsorption (Hiemenz and Rajagopalan 2016). The bi-Langmuir model, in contrast to the traditional Langmuir model, demonstrates temperature sensitivity and provides an accurate depiction of shale's sorption behaviour across a broad temperature range (Lu et al. 1995; Fianu et al. 2018). The traditional Langmuir model, however, falls short when it comes to higher pressures, as it fails to capture the intricate temperature-related intricacies observed in the of shale samples' sorption behavior. Specifically, at elevated pressures, the Langmuir model inadequately incorporates the impact of temperature on shale sample sorption patterns (Lu et al. 1995).

3.2.2 Brunauer–Emmett–Teller (BET) Model

When the reservoir pressure is high, gas is anticipated to form multiple layers on the adsorbent surface, classifying the adsorption isotherm as type II. The BET model was created for molecules that were adsorbed in multiple layers. It serves as a broadened version of the Langmuir model, providing a comprehensive understanding of the adsorption process in situations where multilayer adsorption occurs.

However, the applicability of the BET isotherm diminishes when modelling methane sorption in reservoir conditions, primarily due to the usual occurrence of temperatures surpassing methane's critical temperature. To overcome this constraint, a new parameter known as the pseudo-saturation pressure (P_s) was introduced by

Clarkson et al. (1997), specifically for methane adsorption under elevated temperature and pressure conditions. In such scenarios, P_s effectively functions as the saturation pressure (P_0). For the adsorption of supercritical methane, P_s is determined using the following equation (Yu et al. 2016).

$$\ln(P_s) = 7.7437 - \frac{1306.548}{19.4362 + T} \tag{3.5}$$

where P_s is measured in megapascals, and T is expressed in kelvin.

Equation (3.6) presents the traditional expression of the BET isotherm, assuming an infinite number of adsorption layers

$$\frac{1}{V\left(\frac{P}{P_0} - 1\right)} = \frac{1}{V_m C} + \frac{C-1}{V_m C}\left(\frac{P}{P_0}\right) \tag{3.6}$$

In this case (Eq. 3.6), the relative pressure is denoted by P/P0, the equivalent volume of adsorption is denoted by V, the monolayer adsorption capacity, is denoted by V_m and corresponds to point B in Fig. 3.1, while c represents the energy constant characteristic to the BET equation. It's important to remember that adsorbed layers in shale nanopores are constrained by the finite pore volume, requiring a finite number of layers. Equation (3.7) represents the BET isotherm, which is customised for n adsorption layers (Yu et al. 2016).

$$V = \frac{V_m c \frac{P}{P_0}}{1 - \frac{P}{P_0}}\left[\frac{1 - (n+1)\left(\frac{P}{P_0}\right)^n + n\left(\frac{P}{P_0}\right)^{n+1}}{1 + (c-1)\frac{P}{P_0} - c\left(\frac{P}{P_0}\right)^{n+1}}\right] \tag{3.7}$$

In the scenario where n equals 1, Eq. (3.7) simplifies to the Langmuir model. As n approaches infinity, Eq. (3.7) transforms into Eq. (3.6).

When considering the saturation pressure (P_0) as P_s, Eq. (3.7) effectively simulates the adsorption of methane on shale at elevated temperatures and pressures. BET isotherm used in the adsorption of supercritical methane necessitates the inclusion of three fitting parameters: V_m, c, and n.

3.2.3 Barrett-Joyner-Halenda (BJH) Model

The BJH method is extensively used for determining pore size distribution and pore volumes in shale and coal samples. This technique relies on the capillary condensation theory, employing the Kelvin equation. The approach supposes a cylindrical pore geometry and utilizes desorption branch data from the isotherm (Barrett et al. 1951). The generalized form of the Kelvin equation can be stated as follows (Pirngruber 2016):

$$\ln\frac{P_{cap}}{P_{sat}} = -\frac{V_m\gamma}{RT}\cdot\frac{dA}{dV} \tag{3.8}$$

Pore condensation pressure and pore diameter are correlated, according to the Kelvin equation (Eq. 3.8), where "P_{cap}" denotes capillary pressure, "P_{sat}" is adsorbate saturation, molar volume is represented by V_m, volume change per interface area change is represented by dV/dA, liquid nitrogen surface tension is represented by, the universal gas constant is represented by R, and the temperature (in Kelvin) is represented by T. Assumptions include a general cylindrical pore shape and the absence of interaction between the pore walls and the fluid. The dV/dA metric relies on pore geometry; for spherical pores, dV/dA equals r/2 (with r as the radius), for pores of cylindrical shape, it equals r, and for slit-shaped or elongated pores, it equals 2r or d [where "d" represents the diameter of the more restricted (narrow) dimension]. Through an examination of the pore's geometry, a generalised version of the Kelvin equation can be explained that takes the pore's curvature into account:

$$\ln\frac{P_{cap}}{P_{sat}} = -\frac{2}{r}\frac{V_m\gamma}{RT} \tag{3.9}$$

The Kelvin equation, found in Eq. (3.9), is useful for simulating the forces of capillary working in a pore when the adsorbate condenses and evaporates in low-pressure gas adsorption experiments. In discussions concerning partially filled pore spaces, this becomes particularly crucial since the condensed adsorbate creates a liquid film that adheres to the surfaces of the pores. In such scenarios, the condensed liquid that makes up the liquid film has its vapour pressure lowered by capillary forces within the pore. As a result, there is greater pressure on the concave than the convex side of the interface in the middle of the pore between the gaseous adsorbate and the liquid film. This differential pressure counters surface tension (γ), causing a progressive collapse of the liquid film interface toward the pore wall as pressure drops (see Fig. 3.2). Because of the combined effect of capillary action, capillary condensation takes place at a pressure lower than that required for traditional condensation because the liquid film's pressure is lower than the pressure of the gas inside the pore. In the desorption process, as pressure decreases incrementally (e.g., from step n−1 to step n, with pressure decreasing from Pn−1 to Pn), capillary evaporation takes place within the pore according to the geometry of the pore, as explained by Eq. (3.10).

$$V = \frac{(r_k+\Delta t)^2}{r_p^2}\cdot V_p \tag{3.10}$$

Here, ΔV stands for the amount of desorbed adsorbate during the pressure change step from P_n-1 to P_n, and V_p stands for the exact pore volume. Throughout the desorption process, the ΔV value is influenced by the contractions in the liquid film

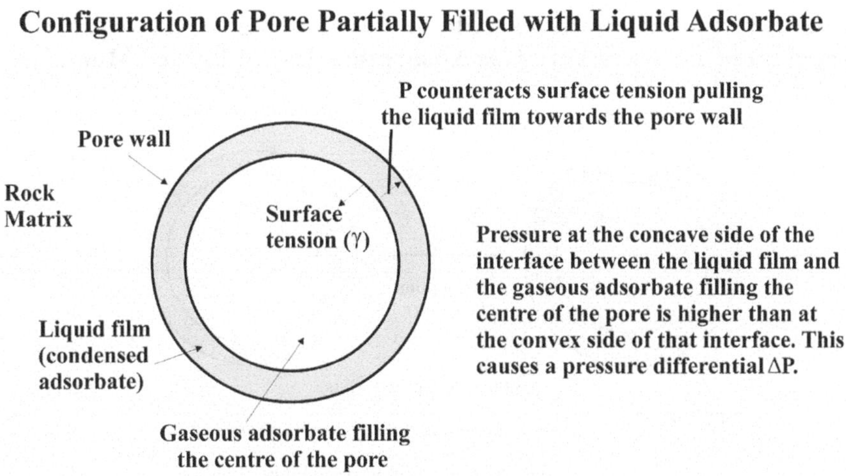

Fig. 3.2 Illustration of capillary forces applied to the liquid film along the pore walls within a cylindrical pore. Reprinted from Springer Nature (Springer ebook), Hazra et al. (2019a), Organic and Inorganic Porosity, and Controls of Hydrocarbon Storage in Shales, Copyright (2019), with permission from Springer Nature

thickness within partially emptied pore sizes. The thickness of adsorbate liquid film surrounding the pore walls is represented by Δt, the change in liquid film thickness during the pressure change step from Pn−1 to Pn, the actual pore radius is indicated by r_p, the capillary radius is defined by r_k, and the thickness of the adsorbate liquid film is indicated by t.

Equation (3.11) describes the simple correlation between rp, rk, and t. Figure 3.3 shows the dynamic interaction between rp, rk, and Δt during the desorption pressure steps.

$$r_p = r_k + t \tag{3.11}$$

By leveraging the capillary effect at different pressure differentials along the desorption isotherm, one can effectively estimate the distribution of pore size of a sample, as illustrated in Fig. 3.4. In each desorption step, the computation of the pore's average diameter experiencing capillary evaporation (r_k) can be achieved by utilizing Eqs. (3.9) and (3.11), leading to the derivation of Eq. (3.12), as long as the true pore radius (r_p) is established.

$$\log(P/P_0) = \frac{-2\gamma V_m}{8.316 * 10^7 * 2.303 T r_k} = \frac{-4.14}{r_k} \tag{3.12}$$

In the specified context, where γ denotes liquid nitrogen's surface tension, V_m represents the nitrogen's molar volume, capillary radius is denoted by r_k, which is usually converted from cm to nm or Å steps, T is the absolute temperature of adsorbate in Kelvin, and the universal gas constant ($R = 8.316 \times 10^7$) is expressed in ergs per degree, the BJH method entails the iterative calculation of pore volume

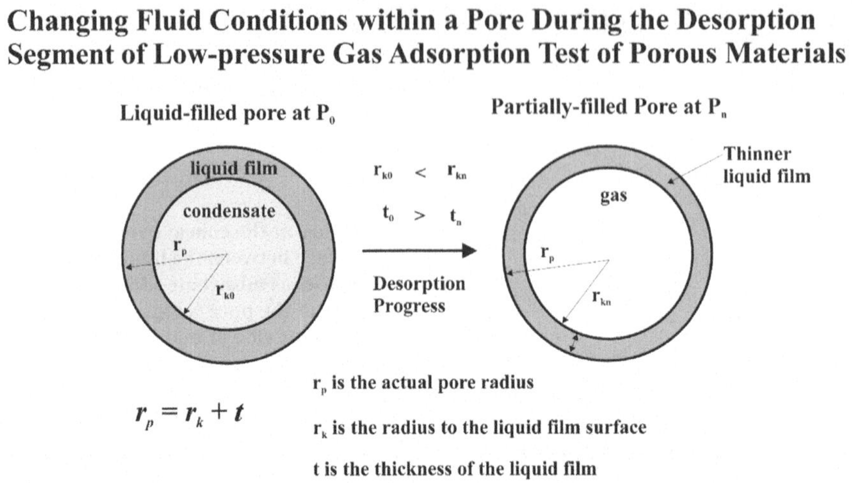

Fig. 3.3 Diagram depicting the correlation between the radius (rk) of the adsorbate gas-filled pore component and the actual pore radius (rp). Reprinted from Springer Nature (Springer ebook), Hazra et al. (2019a), Organic and Inorganic Porosity, and Controls of Hydrocarbon Storage in Shales, Copyright (2019), with permission from Springer Nature

(V_{pn}) across n desorption steps. Equation (3.13) illustrates how it improves upon the capillary evaporation equation by adding a term that accounts for the liquid film's gradually varying thickness with each increment of pressure change.

$$V_{pn} = \frac{r^2}{(r_{kn} + \Delta t_n)^2}.\Delta V_n - \frac{r^2}{(r_{kn} + \Delta t_n)^2}.\Delta t_n.\sum_{j=1}^{j=n-1} \frac{r_{pj} - t_j}{r_{pj}}.A_{pj} \tag{3.13}$$

Here, Δt_n stands for the change in the thickness of the layer in each of the n desorption steps, ΔV_n for the volume desorbed in each one of the n desorption steps, A_p for the surface area of each pore, and r_k for each pressure step is determined by Eq. (3.12). Equation (3.14), which presents a geometric relationship, can be used to calculate Ap, a constant for pores of a given size, for the pressure increase during the desorption process.

$$A_p = 2\frac{V_p}{r_p} \tag{3.14}$$

This implies that Ap can be cumulatively computed for every stage that the desorption process advances to (Barrett et al. 1951). The BJH technique can be applied successfully to the adsorption branch of the isotherm, even though it was initially developed keeping the desorption branch in mind. Significantly, some cases present artificial peaks or steps in the desorption branch at particular pore diameters because of relationships between surface tension and meniscus radius. This compromise in

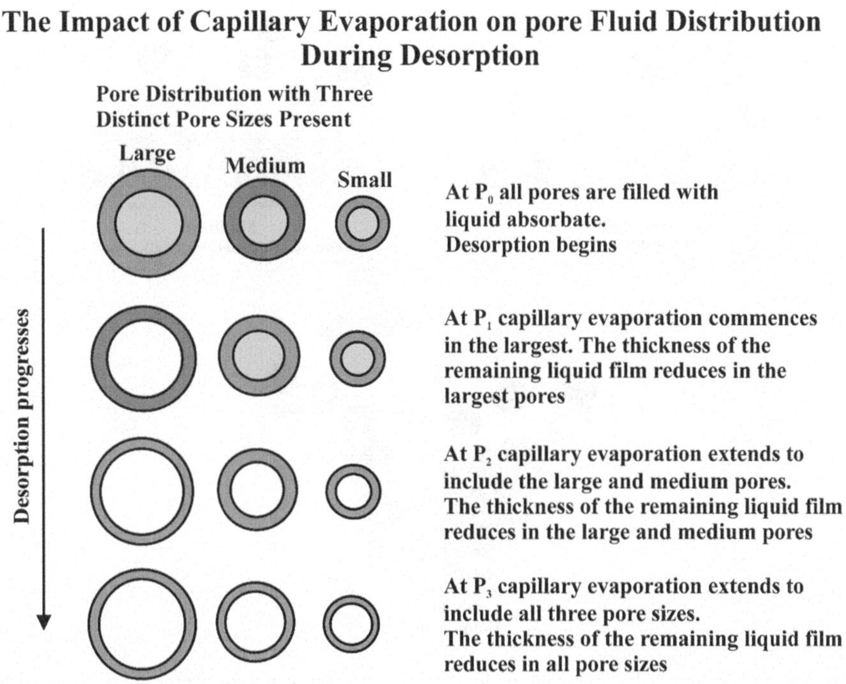

Fig. 3.4 Illustration of the pore-size distribution model as conceptualized by the BJH method. Reprinted from Springer Nature (Springer ebook), Hazra et al. (2019a), Organic and Inorganic Porosity, and Controls of Hydrocarbon Storage in Shales, Copyright (2019), with permission from Springer Nature

the accuracy of calculation of pore volume distribution prompts a more meaningful approach, which entails using the adsorption branch isotherm that is smoother there by enabling precise calculations.

3.2.4 Dubinin's Model

According to the kinetic theory, both the BET and Langmuir models assume surfaces with uniform energy levels, meaning all adsorption sites possess identical adsorption potential (Brunauer 1961). In this framework, the initial adsorbed layer uniformly covers the entire surface before subsequent layers form (Fig. 3.5). However, natural organic shale surfaces exhibit imperfections, containing heteroatoms that lead to variations in adsorption potential across different surface locations. These heterogeneities challenge the theory of energetically homogeneous surfaces made by the Langmuir and BET models.

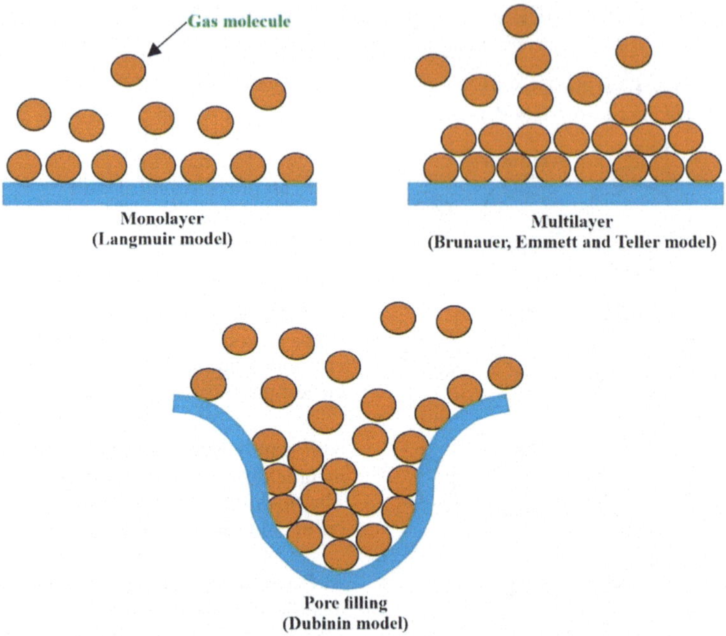

Fig. 3.5 The theoretical framework includes the Langmuir isotherm representing monolayer adsorption, the BET isotherm depicting multilayer adsorption, and the Dubinin model illustrating pore filling. Reprinted from Elsevier Book, titled Coal and Coalbed Gas, Flores (2014), Chapter 4—Coalification, Gasification, and Gas Storage, Copyright (2014), with permission from Elsevier

Rather than immediately attaining total surface coverage, adsorption is thought to start at locations with a high potential for adsorption. This means that molecules that are adsorbed tend to create multiple layers on the surface prior to eventually achieving monolayer coverage, reflecting the complexity and heterogeneity of the surfaces they interact with. For adsorption on surfaces that are flat or pore surfaces with a large radius, both the BET and Langmuir models remain applicable (Ozdemir 2004). As illustrated in Fig. 3.5, adsorption in small pores, or micro-pores, is fueled by mechanism of pore-filling as opposed to layered coverage of the surface (Hutson and Yang 1997). The first pore-filling model is based on potential theory of adsorption by Polanyi (Polanyi 1932), wherein the energy necessary to transfer an adsorbate molecule from the gas phase to the adsorbed phase while being constrained by adsorption forces is known as the adsorption potential. Dubinin interpreted the adsorption potential formulated by Polanyi as the differential molar work (Yang and Liu 2020). They developed the Dubinin and Radushkevitch (D–R) model (Dubinin 1947) and Dubinin-Astakhov (D–A) model (Dubinin and Astakhov 1971) through a approach of thermodynamics. The form of the D–A model is given by

$$V = V_0 \exp\left\{-\left(\left(\frac{RT}{\beta E_0} ln\left(\frac{P}{P_0}\right)\right)\right)^n\right\} \tag{3.15}$$

where V is the volume filled by molecules that are absorbed at P/P_0 relative pressure and V_0 is the overall micropore volume. R is the constant for all gases. The temperature is T. The reference vapor's characteristic energy constant is denoted by E_0. The similarity constant, denoted by β, is equal to 1 in the reference case. N is a tiny integer associated with the adsorption potential distribution. The D–A model simplifies to the D–R model when n = 2.

While shale encompasses a diverse range of pore sizes, the predominant influence on the adsorptive characteristics of the rock comes from micropores, which contribute the highest surface area. Among the various models employed to analyse the adsorption characterstics of microporous solids in shale gas reservoirs, the BET model, Langmuir model, and Dubinin–Polanyi model stand out as the most commonly utilized adsorption isotherms.

3.2.5 Density Functional Theory (DFT) Model

The phenomena of sorption of molecules within a shale or coal matrix can be explained through quantum mechanics (Yu et al. 2020). Quantum mechanics (QM) furnishes a comprehensive insight into the physical and chemical interactions governing the sorption of gas within shale or coal systems, employing a multi-body Schrödinger equation approximation (Eq. 3.16), as noted by Odoh et al. 2015). Within the realm of the QM approach, one method, namely Density Functional Theory (DFT) (Odoh et al. 2015), is employed to simulate adsorption properties of gas at a macroscopic level, offering a quantum chemical portrayal of the adsorbate-adsorbent relationship, as highlighted by Wang et al. (2019). The DFT model, grounded in the Schrödinger equation, is expressed as follows.

$$H\Psi = E\Psi \tag{3.16}$$

In this context, where E represents energy of the system, H denotes the Hamiltonian, and the wave function is represented as Ψ, the application of the DFT model has been embraced for the examination of gas adsorption and interactions within shale systems, as documented by Wang et al. in (2018). Equation (3.17) serves the purpose of computing the adsorption or desorption energy at an adsorption site through the utilization of the DFT model.

$$A_e = E_{adsorbate} + E_{host} - E_{adsorbate/host} \tag{3.17}$$

In this equation, Ae represents the energy associated with adsorption or desorption, $E_{adsorbate}$ stands for the energy of the single-point of the rock matrix, E_{host} denotes energy of gas molecule, and $E_{adsorbate/host}$ corresponds to the of the matrix-gas system's single-point energy.

3.2.6 Grand Canonical Monte Carlo (GCMC) Method

Monte Carlo simulation is widely employed for investigating the confinement effects in nanopores. This technique achieves properties of thermodynamics and the system structure with the maximum probability through continued sampling and generation of molecular system configurations, followed by the calculation of sum of their energy (Liu and Zhang 2019). The GCMC simulation is used widely to simulate fluid behaviours and properties in order to study adsorption. System energies in GCMC simulations are iteratively gotten until the individual species' chemical potentials align with specified values. By introducing, eliminating, or rotating molecules at random, this is accomplished (Liu and Zhang 2019). GCMC involves the computation of temperature, chemical potential, and volume, typically keeping these factors constant while varying the energy and particle numbers. The equation of state is applied for the straightforward correlation of fugacity, an essential parameter as an input to the GCMC model, with chemical potential (Wang et al. 2018). Song et al. (2018) in their study evaluated adsorption of methane in shale systems affected by its pore structure using the GCMC technique. Their results demonstrated that adsorption capacity is highly dependent on the types of pores, highlighting the significance of characterization of pore structure in the estimation of capacity of adsorption in shale formations. Similarly, Mosher et al. (2013) evaluated methane adsorption throughout different pore sizes, revealing a substantial influence of pore size on adsorption in shale formations. However, they observed that the molecular simulation method have a tendency to underestimate adsorption in excess at pressures under 2.5 MPa while overestimating at pressures beyond that. The discrepancies arose from limitations inherent in the models employed to simulate pore sizes in the study.

3.2.7 Molecular Dynamics (MD) Simulation

MD simulations emulate real atoms by resolving Newton's equation of motion, enabling the calculation of phase trajectories for molecular systems. Particularly well-suited for fluid flow studies, MD simulations have found application in exploring the adsorption mechanisms within shale systems. They serve as a valuable alternative to actual experiments, confirming or validating conceptual frameworks and uncovering novel phenomena (Liu and Zhang 2019). Despite the profound insights offered by MD simulations into system behavior, their computational expense is significant,

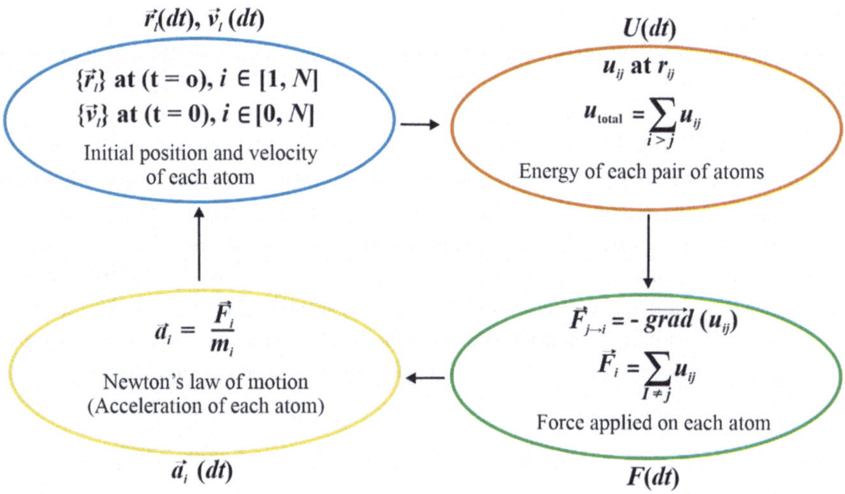

Fig. 3.6 In the molecular dynamics cycle, "U" denotes energy, "r" represents the distance between two atoms, "a" represents acceleration, "F" stands for force, and "v" signifies velocity. Reprinted from Journal of Petroleum Science and Engineering, Vol. 208, Babatunde et al. (2022), Adsorption of gases on heterogeneous shale surfaces: A review, Pages No. 109466, Copyright (2022), with permission from Elsevier

and time steps are limited. During each iteration, the computational cycle illustrated in Fig. 3.6 is carried out. Commencing an cycle of MD requires knowledge of the preliminary configuration of atoms, encompassing their initial positions and velocities at time zero. This arrangement is then utilized to compute the systems' energy. The resultant force, a summation of forces within the system derived from the differentiation of energy for each atom pair, is subjected to Newton's law of motion to determine atomic acceleration. This information, coupled with the initial system configuration, predicts the new positions of the atoms through numerical integration.

3.3 Adsorption Methods

3.3.1 Low Pressure Gas Adsorption

Low-pressure gas adsorption (LPGA) analysis has been employed for several years to study the porous structures within coals and shales. Nitrogen is frequently employed as the gas of choice in low-temperature and low-pressure gas adsorption experiments performed on shales or coals. Samples with pore sizes ranging between 1.7 and < 200 nm can be studied by using cryogenic liquid nitrogen maintained at a −197.3 °C constant temperature condition and relative pressure range of almost zero to 1.0 (Kuila and Prasad 2013). Nevertheless, one drawback of nitrogen gas use is that not all macropores and micropores with pore widths larger than 200 nm can be reached. Many researchers have used CO_2 for micropore evaluation, which typically involves low pressures and 0 °C temperatures (Ross and Bustin 2009; Mastalerz et al. 2012). In LPGA analysis, nitrogen is used as the adsorbate, and the temperature is kept constant. P/P0, or relative pressure levels, are systematically increased; P stands for gas vapour pressure, and P0 for the adsorbent's saturation pressure. This aids in the construction of the adsorption isotherms by providing the volume of adsorbed gas at various P/P0 intervals. The adsorption isotherm produced provides a qualitative assessment of the sample's pore structure (Brunauer et al. 1938). Predicting the pore types found in the rock is made easier by the hysteresis pattern and the shape of the adsorption isotherms. In accordance with IUPAC, there are six different types of adsorption isotherms (Fig. 3.1) and four different types of hysteresis patterns (Fig. 3.7).

3.3.2 *High Pressure Gas Adsorption*

Manometric, volumetric, and gravimetric techniques find extensive application in high-pressure sorption studies on coal and shale (Mudoi et al. 2022). While the gravimetric method, based on mass, offers high accuracy, its applicability is constrained by small sample sizes. Conversely, the volumetric method, grounded in Boyle's law, demonstrates suitability for highly heterogeneous samples such as coal and shale rocks. The initial step in the adsorption experiment involves accurately estimating the adsorbent's void volume. For the volumetric and manometric methods, precise void volume estimation is crucial for constructing error-free isotherms (Busch and Gensterblum 2011).

3.3.2.1 Monometric Method

The manometric technique involves the sequential transfer of a predefined volume of gas from a reference volume (calibrated) into the sample cell that holds the specimen being examined (Fig. 3.8). The sample cell's void volume (V_{Void}^{0}) is measured through the increase in a "non-sorbing" gas volume, usually helium, before the sorption experiment is started. Helium densities are computed through the application of the van der Waals equation, utilizing the parameters a and b as proposed by Michels and Wouters (1941), or by employing the equation of state (EOS) developed by McCarty and Arp (1990). This process also determines the skeletal density (V_{sample}^{0}) and skeletal volume (V_{sample}^{0}) of the sample. For gas sorption isotherms, the "non-sorption" reference mass is obtained by multiplying the void volume with gas density (or supercritical) phase ($V_{Void}^{0}\rho^{CO_2}(T, p)$). In the event that there was no sorption,

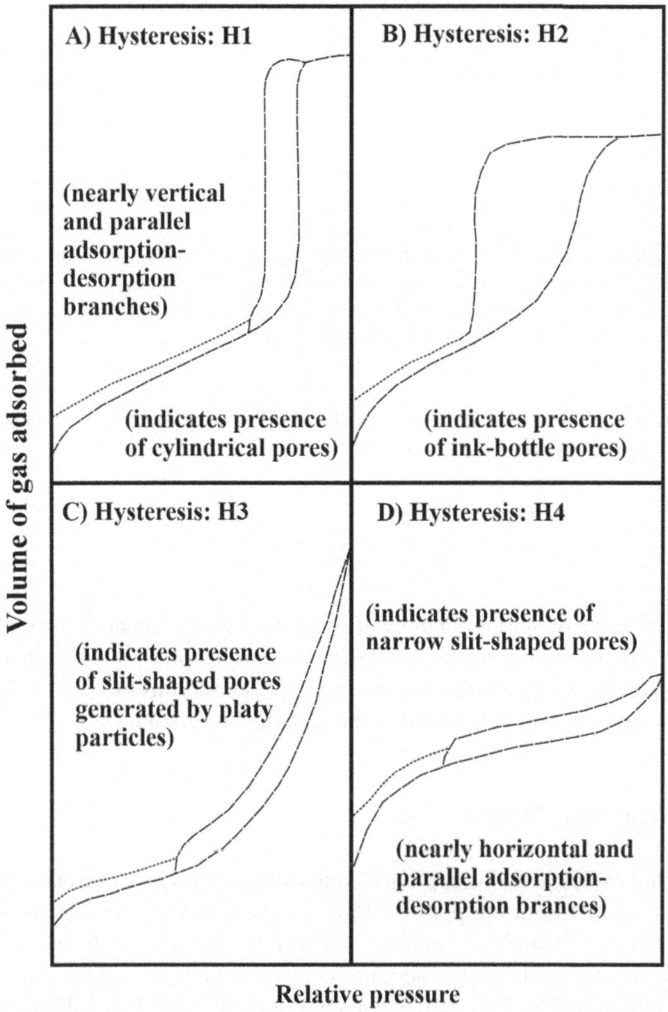

Fig. 3.7 Different hysteresis pattern as given by IUPAC (Sing 1985; Pure and applied chemistry, 57(4), pp. 603–619. Copyright © 1985 IUPAC)

this amount would be the gas (supercritical fluid) that filled the measurement cell. Densities are determined through the utilization of the respective equation of state (EOS) for N_2, CH_4, CO_2, or their combinations of binary and ternary mixtures. The excess sorption mass ($m^{CO_2}_{excess}$) is determined by subtracting the "non-sorption" reference mass from the mass of gas introduced into the measuring cell up to a designated pressure step:

$$m^{gas}_{excess} = m^{gas}_{transferred} - V^0_{void} \cdot \rho^{gas}(T, p) \tag{3.18}$$

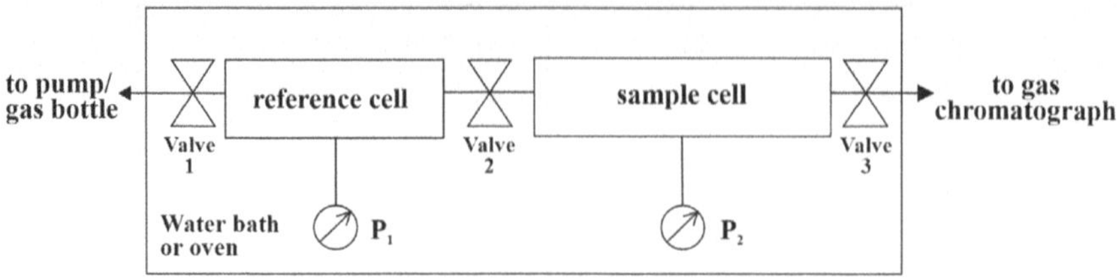

Fig. 3.8 Diagram depicting of the setup for high-pressure isotherm measurement through manometric method. Reprinted from International Journal of Coal Geology, Vol. 87, Busch and Gensterblum (2011), CBM and CO_2-ECBM related sorption processes in coal: a review, Pages No. 49–71, Copyright (2011), with permission from Elsevier

The transfer of mass to the measuring cell from the reference cell over N successive pressure steps is expressed as:

$$m^{gas}_{transferred} = \sum_{i=1}^{N} V_{ref}.\left(\rho^{gas}_{i}(p_i T) - \rho^{gas}_{i-1}(p_{i-1}, T)\right) \tag{3.19}$$

Instead of being quantified in units of mass, the surplus quantity of sorbed gas is commonly articulated in terms of substance amounts (mol) or the volume of sorbed gas under standard temperature and pressure (STP) conditions (m^3; 288 K, 0.1 MPa). The excess sorption is then adjusted to the sorbent's original mass.

3.3.2.2 Volumetric Method

As previously mentioned, this approach adheres to Boyle's law, mirroring the procedure employed in measuring porosity through pycnometry. The system comprises two compartments, namely one sample cell and one reference cell, segregated by a valve, both situated within a temperature-controlled bath (Fig. 3.9). The reference cell volume, denoted as VR, is a established quantity, which is calibrated against the volume of aluminium plugs (these plugs are precisely machines for accuracy). Following the introduction of the crushed particle into the sample cell, helium expansion in the reference cell are used to calculate the void volume, or Vs, which includes the space between particles, within lines, and within the porosity of the particles. The quantification involves assessing the quantity of molecules introduced into the cell in the following manner:

$$n_{total} = \frac{P_1 V_R}{Z_1 R T_1} \tag{3.20}$$

In the provided equation, V_R represents reference cell volume, Z_1 signifies the factor of compressibility, P_1 denotes the pressure, R stands for the universal gas constant, and T1 corresponds to the temperature, specifically the oil bath temperature

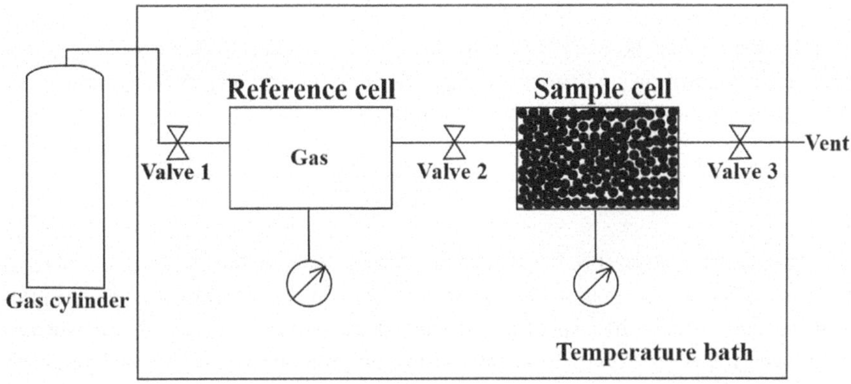

Fig. 3.9 Diagram depicting the setup for high-pressure isotherm measurement through volumetric method. Reprinted from Journal of Unconventional Oil and Gas Resources, Vol. 8, Heller and Zoback (2014), Adsorption of methane and carbon dioxide on gas shale and pure mineral samples, Pages No. 14–24, Copyright (2014), with permission from Elsevier

that is kept at 40 °C. Subsequently, the valve, which acts as a separator for the sample and the reference cells, is opened, permitting the gas to occupy the sample cell's void volume. Filling the void volume with what is presumed to be a non-adsorbing gas (helium) is the reason for the subsequent pressure drop that occurs when the valve is opened. Consequently, the following equation can be used to calculate the void volume of the sample cell because the total amount of molecules that are free in the system stays constant:

$$V_s = \frac{n_{total} Z_2 R T_2}{P_2} - V_R \tag{3.21}$$

In the above equation (Eq. 3.21), P_2 denotes the pressure, Z_2 represents factor of compressibility, and T_2 represent the equilibrium temperature after the opening of the valve. After measuring the volume of void using helium, the system is vacuumed to remove any remaining helium that might have become trapped in the lines or specimen particles. Once the reference cell void volume is measured, subsequent increase of the gas adsorbed in the specimen can be accomplished utilizing the same approach as previously described. On the other hand, void volume filling and adsorption cause pressure drops as an adsorbing gas expands and moves from the reference cell to it. Notably, molecules that have adsorbed will certainly have no kinetic energy, therefore do not contribute to the gas phase pressure.

Consequently, the adsorption amount is then measured using the following formula:

$$n_{ads}^{excess} = n_{total} - n_{free} \tag{3.22}$$

In the provided equation (Eq. 3.22), n_{total} represents the initially expected quantity of moles present in the reference cell (the moles anticipated in the system in the absence of adsorption). Meanwhile, n_{free} signifies the initially present number of moles at equilibrium in the free phase, computed as follows:

$$n_{free} = \frac{P_2(V_R + V_S)}{RT_2Z_2} \tag{3.23}$$

In the given equation (Eq. 3.23), V_R and V_S denote the volumes of the sample and reference cells, respectively., The valve dividing the two chambers is sealed after measurement is taken at a specific pressure, then the reference cell is pressurized further, and the entire process is iterated till a comprehensive isotherm is delineated.

3.3.2.3 Gravimetric Method

Gravimetric systems track the mass variation in the sample during the process of gas adsorption. By directly comparing it with the mass increase of an empty chamber upon pressurization, issues related to equations of state for estimating the non-adsorbed gas are circumvented. Gravimetric methods are particularly well-suited for applications involving supercritical CO_2 or mixed gas adsorption. The schematic in the Fig. 3.10 illustrates a gravimetric experimental setup. In this apparatus, the sorbent is positioned within a compartment that is at high pressure and is connected to a magnetic suspension balance while being subjected to sorptive at constant temperature. The surplus is sorption is deduced from the difference in weight (apparent mass change) of the specimen (Eq. 3.24) which is recorded at the time of distinct sorption steps, where m^0_{sample} denotes the initial mass of the sample.

$$\Delta = m_{measurement}(T, p) - m^0_{sample} \tag{3.24}$$

The surplus sorption is obtained through the change in apparent weight, with a correction of buoyancy reliant on the skeletal volume (V^0_{sample}) of the specimen, aligning with the reference state akin to the method of manometric method. Helium is utilized for the measurement of the volume or density of the skeleton. The excess sorbed mass is expressed as:

$$m^{gas}_{excess} = \Delta m + V^0_{sample}\rho_{gas}(T, p) \tag{3.25}$$

or

$$m^{gas}_{excess} = \Delta m + \frac{m^0_{sample}}{\rho^0_{sample}}\rho_{gas}(T, p) \tag{3.26}$$

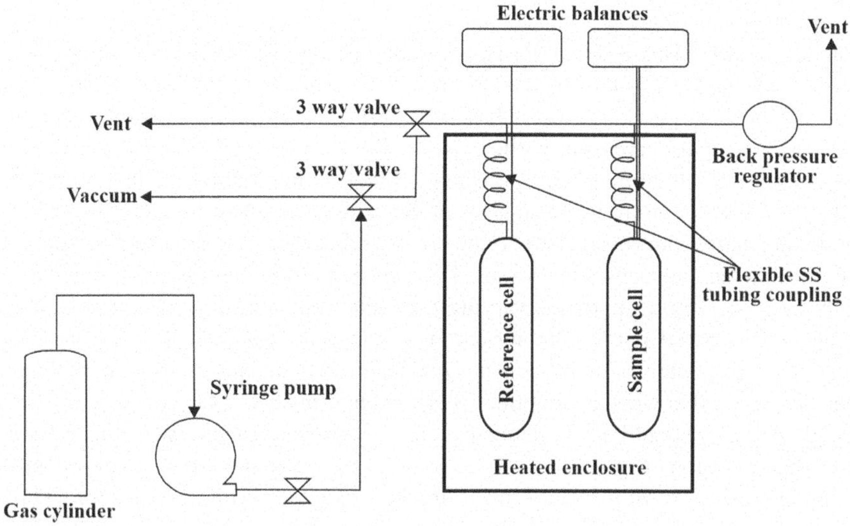

Fig. 3.10 Diagram depicting the setup for high-pressure isotherm measurement through gravimetric method. Reprinted from Elsevier Book, titled Greenhouse Gas Control Technologies 7, Day et al. (2005), Factors controlling CO_2 absorption in Australian coals, Volume II – Part 2, Copyright (2005), Pages 2221–2224, with permission from Elsevier

Similar to the procedure of manometric method, the density and volume of the phase of sorption remain unknown, making it challenging to explicitly factor their impact into the buoyancy term. The values of excess sorption are then adjusted to the original mass of the sample (m^0_{sample}).

3.4 Factors Influencing LPGA Studies

3.4.1 Impact of Particle Crush Size

For optimal low-pressure gas adsorption (LPGA) analysis studies, it is preferable to crush samples into finer sizes. This approach minimizes the path-length, facilitating improved access of gases to the intricate structures of pores within the sample (Kuila and Prasad 2013). Studies on shale-core crush-rock methodology was initially introduced by the Gas Research Institute (GRI) (Luffel and Guidry 1992). Due to the absence of universally agreed-upon standard crush size, various studies have employed various sizes for low-pressure gas analysis experiments, such as 8 mm (Schmitt et al. 2013), 1.4 mm to 250 microns (Mohammad et al. 2013), 1 mm to 800 microns (Hazra et al. 2018a), and less than 250 microns (prevalently employed;

Ross and Bustin 2009; Chalmers et al. 2012; Clarkson et al. 2013; Yang et al. 2014; Wang et al. 2016; Hazra et al. 2018a). Many studies have shown the effect of particle crush-sizes on the results obtained from LPGA analysis (Chen et al. 2015; Han et al. 2016; Wei et al. 2016; Mastalerz et al. 2017; Hazra et al. 2018b). Chen et al. (2015) found that decreasing particle crush size in experiments of gas adsorption led to increased volumes of mesopores, as volumes of micropores exhibited varying outcomes. Reduction in the crush sizes of the samples resulted in tighter hysteresis loops, indicating improved pore connectivity and earlier attainment of adsorption equilibrium. In another study on shale samples from Langmaxi Formation (China) with sizes of samples particles varying between from 4 mm to 58 microns, Han et al. (2016) noted rise in the surface area and pore volumes, accompanied by tighter hysteresis loops, as the crush sizes decreased. According to their findings, they suggested that when conducting LPGA studies on shales, the crush sizes of the sample particles should be maintained < 113 microns. In contrast, Wei et al. (2016), from their research on the shale samples from Sichaun Basin (China), recommended utilizing crush-sizes of the samples varying between 250 to 140 microns for LPGA experiments. In their research on the shale samples (with varying levels of thermal maturity) from Illionois Basin, Mastalerz et al. (2017), suggested employing sample sizes of 75 microns for LPGA experiments.

In line with the previous researchers, Hazra et al. (2018b) noted marked variations in the N_2 and CO_2 adsorption results when altering crush sizes of sample ranging between 1 mm to 53 microns were used. They studied two shale samples of contrasting thermal maturities: one oil-window mature shale and another overmature shale. Each sample was carefully and slowly crushed with the help of pestle mortar to four different sample sizes for their study: (a) size S1: 1 mm–500 microns; (b) size S2: 500–212 microns; (c) size S3: 212–75 microns; and (d) size S4: 75–53 microns. As shown in Fig. 3.11, the shale sample at oil-window maturity showed a raise in adsorption capacity that was consistent with a decrease in particle size (S4 > S3 > S2 > S1) (Fig. 3.11a–d). The Type IV isotherm shape exhibited by particle sizes S1 and S2 is defined by the absence of steep slopes in the 0.98–1.00 relative pressure range, which suggests the presence of mesopores and the absence of a significant number of macropores. However, as the particle size decreases even more (S3 and S4), the slopes gradually increase within the 0.98–1.00 relative pressure range, which aligns with an increase in the capacity for nitrogen sorption. This finding also points to a higher number of macropores in the S3 and S4particle sizes (Fig. 3.3c, d). This was attributed to nitrogen gas's enhanced ability to access pores when particle sizes are smaller, potentially alleviating prior constrictions imposed by the shale matrix. With the decrease in the size of the particles from S1 to S4, there is a noticeable trend of tightening desorption hysteresis loops (Fig. 3.3a–e). The profile of isotherm of S3 particle size was characterized as Type IIB, as defined by Rouquerol et al. (2013). This type includes some macropores that explain the lack of a plateau-like structure usual of Type IV isotherms (mesoporous), as well as mesopores that contribute to the observed hysteresis. The S4 size isotherm shape is more similar to isotherm shape of Type II, albeit without complete reversibility and exhibiting some hysteresis, indicative of the mesopores' presence. Notably, all sample sizes exhibited hysteresis loops of type H, pointing towards the existence of pores that are slit-shaped (Gregg et al. 1967; Sing 1985).

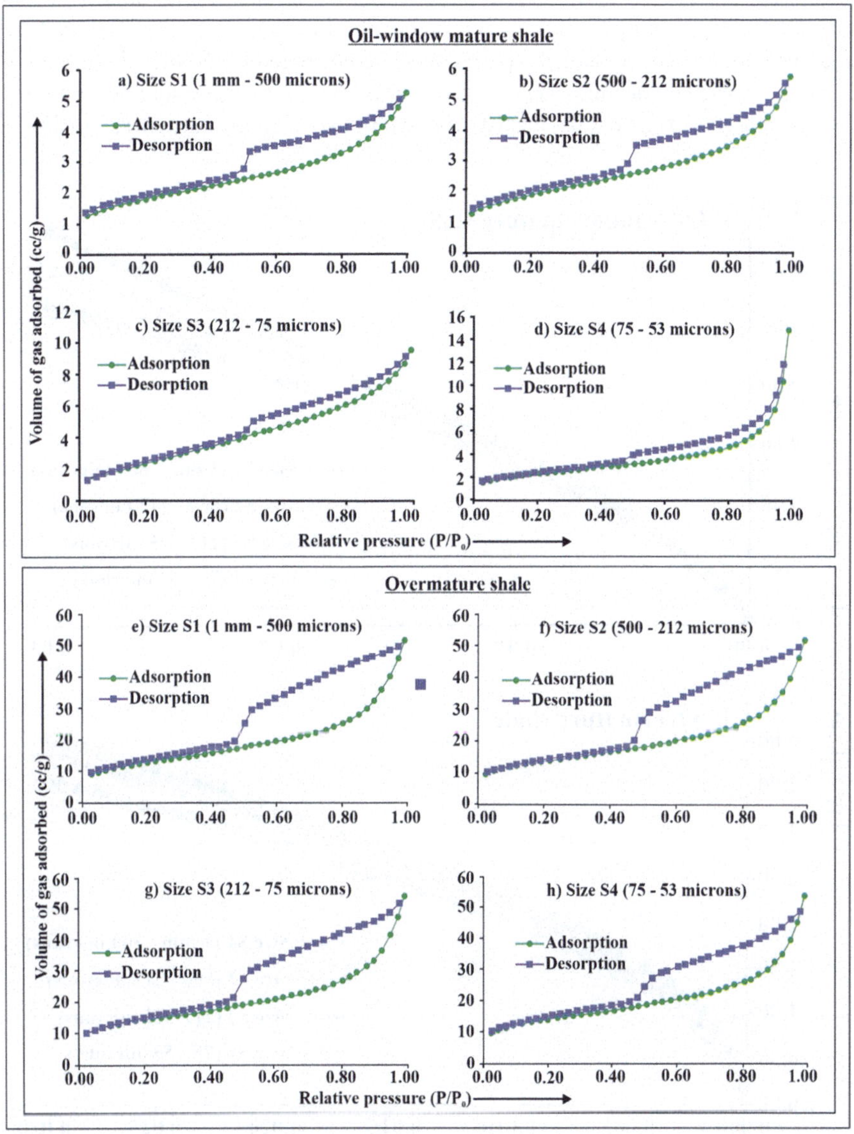

Fig. 3.11 Nitrogen gas adsorption–desorption isotherms of oil-window mature and overmature shale sample. Reprinted (adapted) with permission from Hazra et al. (2018b); Energy Fuels 2018, 32, 8, 8175–8186. Copyright © 2018 American Chemical Society Copyright © 2018 American Chemical Society

The BJH pore volume (cc/g) was noted to rise as crush size of the particles decreased (S4 > S3 > S2 > S1; Table 3.1). The BET specific surface area (SSA) exhibited an ascending trend from particle sizes S1 to S3, followed by a subsequent decline for size S4 (S3 > S4 > S2 > S1). This observed increase in both BET and BJH derived SSA and pore volumes, respectively, for sizes S1 to S3 suggested the emergence of previously isolated pores, which contributed to the overall volume of the pores. However, the unique behavior of size S4, where the pore volume continued to rise while the BET SSA decreased, implied that at such fine particle sizes (75–53

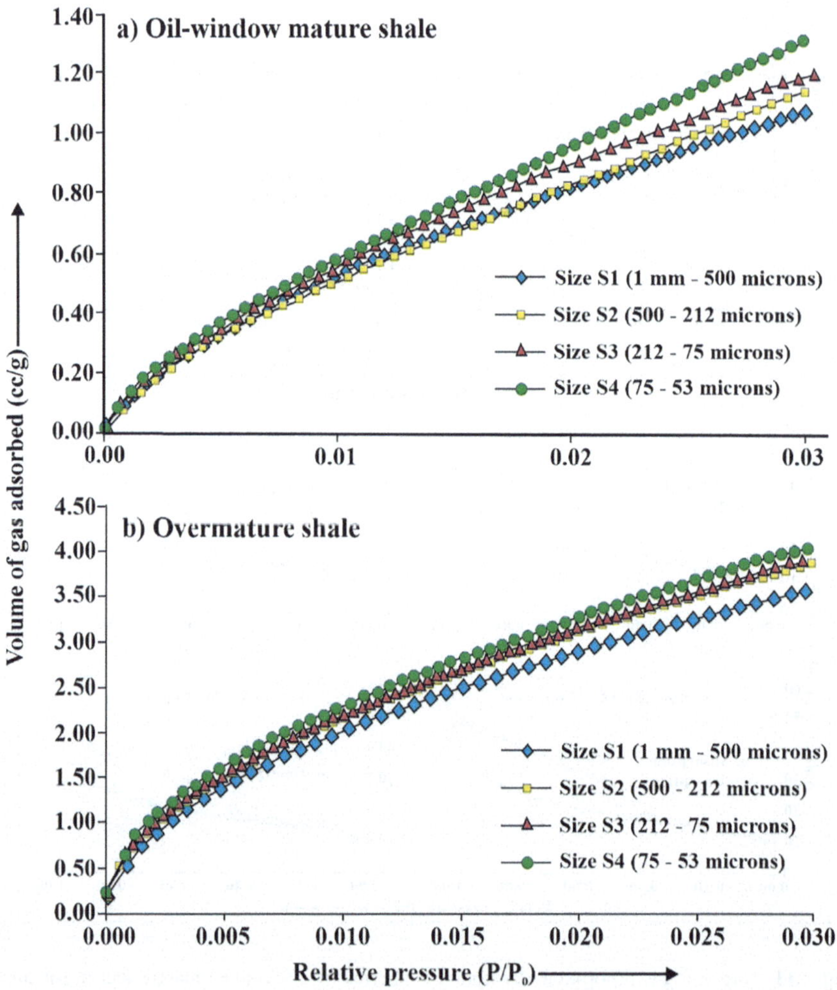

Fig. 3.12 Carbon dioxide gas adsorption isotherms of oil-window mature shale (**a**) and overmature shale (**b**) at four different particle sizes. Reprinted (adapted) with permission from Hazra et al. (2018b); Energy Fuels 2018, 32, 8, 8175–8186. Copyright © 2018 American Chemical Society

microns), the increased pore volume was attributed to larger pores that have minimal impact on the surface area. Contrastingly, the average pore diameter demonstrated

Table 3.1 Pore surface area, pore radius, and pore volume of the oil-window mature and overmature shale samples analysed using crushed specimens with four varying particle sizes.

Sample		S1 (1 mm to 500 microns)	S2 (500–212 microns)	S3 (212–75 microns)	S4 (75–53 microns)
Oil-window mature shale	BET SSA (m^2/g)	6.178	6.427	9.689	7.840
	Total pore volume (cc/g)	0.008	0.009	0.015	0.023
	Average pore radius (Å)	26.19	27.70	30.27	58.48
	V_G (cc/g)	5.23	5.76	9.48	14.82
	BJH Pore volume (cc/g)	0.006	0.007	0.013	0.021
Overmature shale	BET SSA (m^2/g)	44.579	46.566	48.772	48.865
	Total pore volume (cc/g)	0.080	0.080	0.084	0.083
	Average pore radius (Å)	35.78	34.33	34.32	33.58
	V_G (cc/g)	51.56	51.68	54.11	54.04
	BJH Pore volume (cc/g)	0.067	0.065	0.068	0.066

Table 3.2 DR micropore surface area and DA limiting micropores volume of the oil-window mature shale and overmature shale at different particle sizes

Sample	Sample sizes	DR micropore surface area (m^2/g)	DA limiting micropore volume (MPV; cc/g)
Oil-window mature shale	1000–500	8.89	0.0117
	< 500 to 212	8.09	0.0140
	< 212 to 75	8.88	0.0115
	< 75 to 53	10.72	0.0176
Overmature shale	1000–500	25.68	0.0411
	< 500 to 212	27.51	0.0425
	< 212 to 75	29.41	0.0466
	< 75 to 53	28.51	0.0442

an increase with diminishing particle crush size (S4 > S3 > S2 > S1) (Table 3.1). This signifies that the transition from S1 to S3 results in the sequential release and accessibility of a significant number of previously blind or isolated mesopores and a few macropores by nitrogen gas. In addition to increasing surface area, this dual liberation also increases pore volume. However, the shift from S3 to S4 primarily leads to the liberation of a considerable number of previously isolated or blind macropores, potentially altering the mesopore structures. For size S4, this leads to a lower surface area but a higher pore volume than for size S3. Low-pressure CO_2 gas adsorption isotherms of oil-window mature sample for the crush sizes S1 to S3 were observed to alter within the range of 10% (Fig. 3.12a). For the smallest particle size, S4, the DA limiting micropore volume and the DR micropore surface area reached their maximum values. This suggested that reduction in sample particle size led to the exposure and liberation of previously isolated micropores, subsequently enabling their accessibility by CO_2 gas. The overmature shale sample exhibited notably higher values for BET specific total pore volume (cc/g), surface area [SSA (m^2/g)], and total volume of nitrogen gas adsorbed (cc/g) compared to the other sample (Table 3.2). A consistent trend emerged with decrease in the size of the particles (from S1 to S4), with BET SSA showing an incremental rise. Notably, for S4 and S3 particle size fractions, the BET SSA values were nearly identical at 48.772 and 48.865 m^2/g, respectively. Correspondingly, the adsorbed volume of gas exhibited a similar trend for size fractions S3 and S4. The lack of sharp slopes in the 0.98–1.00 relative pressure range across all sizes, or S1–S4, suggested mesopores' presence and the absence of large macropores (Fig. 3.11). The lack of a plateau suggested macropores' presence, categorizing the shape of the isotherms as Type IIB according to Rouquerol et al. (2013). Across all sizes, hysteresis loops of Type H4 were evident, implying the existence of slit shaped pores (Gregg et al. 1967; Sing 1985). While not as pronounced as in the case of oil-window mature shale, the overmature shale sample displayed slightly tightening desorption hysteresis loops with decrease in the particle crush sizes (i.e., S1–S4) (Fig. 3.11e–h). Also, in contrast to the oil-window mature shale, the overmature shale specimen exhibited a reduction in average pore diameter as size of the particles decreased. Although, the difference in average pore diameter among various sizes of particles in the overmature shale sample is considerably less pronounced compared to the oil-window mature sample. Furthermore, the overmature shale sample's nearly constant total pore volume was noted for each of the four particle crush sizes (Table 3.1) suggesting that unlike the oil-window mature shale, a substantial share of closed pores or previously isolated pores in overmature shale are already available even at coarser particle size fractions. Figure 3.12b illustrates low-pressure adsorption isotherms derived using CO_2 for four size fractions, all exhibiting Type I characteristics with notably higher adsorbed gas volumes (approximately 3.5 to 4.0 cm3/g STP at 0.03 relative pressure (P/P0)) compared to the oil-window mature sample. A consistent pattern emerges with a rise in CO_2 adsorption capacity as size of the crushed particle decreases (Fig. 3.12). The DR micropore surface area (MSA; m^2/g) shows an initial increase for size fractions S1–S3, followed by a minor decrease for the particle size fraction S4 (MSA ranking as S3 > S4 > S2 > S1, Table 3.2). Similarly, the DA limiting micropore volume (MPV; cc/g) exhibits an ascent from S1–S3, followed by a minor decline for S4 (MPV ranking as S3 > S4 > S2 > S1; Table 3.2). These trends, akin to oil-window mature shale, indicate the exposure and liberation of isolated micropores with decrease in the sizes of the crushed particles, rendering them accessible to CO_2 gas. The findings of Hazra et al. (2018b) confirmed that the particle-crush size significantly affects measurement of pore structure of shale samples, introducing a complex yet noteworthy impact that shapes the interpretation of such data.

3.4.2 Impact of Degassing Time and Temperature

A crucial prerequisite before conducting LPGA experiments involves degassing the specimens to eliminate residual water or gases, ensuring that the adsorbate can effectively enter the diverse porous structures (Holmes et al. 2017). However, just like with size of the crushed particles, no widely endorsed standard exist for the temperature and time at which sample degassing should occur. Holmes et al. (2017) tried to standardise the procedure of degassing of shale and proposed 250 °C as the temperature. While various researchers have employed different degassing times and temperatures, a prevailing choice across multiple studies has been a temperature of degassing of 110 °C for a duration of 12–14 h. Furthermore, clay mineral type found in the specimen can be greatly affected by the length and temperature of the process of degassing. Likewise, at different degassing temperatures, the response of the organic matter (in relation to composition and maturity) and the structure of the kerogen can display variety of results. In accordance with the structure of kerogen models formerly suggested by Behar and Vanderbroucke (1987), it is firmly showed that distinct types of kerogens exhibit varying behavior as they undergo progressive maturation. For instance, Type-I kerogen, characterized by the presence aliphatics that are long-chained (> 40 carbon atoms) in the early phases of the process of diagenesis, produces significant quantities of gas and paraffinic oil with progression of maturity. On the other hand, kerogen belonging to Type-III category exhibits a higher amount of oxygenated functional groups and reduced hydrogen content at the beginning of the process of diagenesis, with a significant portion of carbon sealed in a remaining fragment featuring of polycyclic aromatic structures of high proportion. Therefore, Type-III kerogen has a much higher bond dissociation energy compared to kerogen belonging to Type-I category. This distinct reaction of kerogen to temperature fluctuations could impact porosity alterations based on the chosen degassing conditions.

Singh et al. (2021b) studied the influence of particle crush sizes of sample, degassing temperature, and time on the evolution of organic-rich shales' (of contrasting thermal maturity) porosity. Preceding the LPGA analysis, rock samples were degassed at 110, 200, and 300 °C for durations of 3 and 12 h under vacuum. In the case of low-mature shale (LS), the crush particle size exhibited limited impact on results of nitrogen gas adsorption. Conversely, temperature of degassing significantly influenced pore attributes, while the duration at specific temperatures had minimal impact. The adsorption–desorption isotherms exhibited similar characteristics for both sizes of crushed samples subjected to degassing at the temperature of 110 °C for 3 and 12 h (Fig. 3.13). At 110 °C, the adsorption isotherm displayed a Type IIB pattern, suggesting the existence of irregularly shaped, narrow-mouthed pores with a broad distribution of pore sizes. The observed pattern pore-filling suggested a greater prevalence of mesopores of larger sizes. Additionally, the hysteresis pattern at this temperature was identified as type H2, indicative of capillary condensation, which results from the large number of pores of ink-bottle shape. Elevating the temperature of degassing to 200 °C induced a notable transformation in the adsorption isotherm characteristics. The LS shale sample exhibited a distinct macroporous nature, evident in the higher adsorption capacity at elevated relative pressures, manifesting as a steeper slope, reminiscent of material of microporous nature. A notable rise in both BJH and DFT pore volumes was observed, providing strong indications that degassing the LS shale sample at 200 °C leads to significant changes of the structure within the sample, thereby presenting a distinct pore system signature. On the other hand, it can be posited that the predominantly macroporous nature, as reflected in the adsorption isotherm, might stem from the exclusion of volatiles in the specimen, leading to the macropores' opening, rather than being solely attributed to alterations in the sample's organic matter structure. Insights into potential alterations occurring during the degassing of the LS shale sample at 200 °C is provided by Rock–Eval and thermogravimetric analysis. The existence of hydrogen-rich, reactive kerogen, evident in elevated hydrogen index (HI) and strong Flame Ionization Detector (FID) signals, attributed to the organic matter cracking within the sample at lesser temperatures, yielding increased hydrocarbon production. Degassing at temperatures of 200 °C or higher induces structural changes in the organic composition. Specifically, the 200 °C degassing reveals a distinct macroporous structure, contrasting sharply with the outcome at 110 °C. Further, at the degassing temperature of 300 °C, the sample's organic matter undergoes additional alterations, showing an isotherm resembling degassed sample at 110 °C. Their findings revealed that degassing of the samples at 110 °C, the pore volume was more pronounced in the smaller range of the pore sizes ($<$ 7 nm diameter) in comparison to the samples degassed at 200 °C. At 200 °C, a reduction in the volume of pores within the smaller size range was noted; beyond that, though, a notable raise in the volume of pores with large diameters became evident. Adsorption isotherm data supporting this observation showed that macroporous structures were present at higher relative pressures, as shown by steeper slopes. The combination of Rock–Eval and thermal analysis results effectively indicated structural changes within the LS shale sample, leading to the closure of pores that were smaller in size and the formation of macropores artificially. Upon increasing the temperature of degassing to 300 °C, distinctive features were observed, reinforcing the influence of degassing temperature on porosity estimation. Specifically, at 300 °C, the isotherm shapes resembled those obtained at 110 °C (Fig. 3.13). According to this observation, the macroporous structure that was created during the 200 °C degassing of the LS sample changes into a mesoporous structure during the 300 °C degassing.

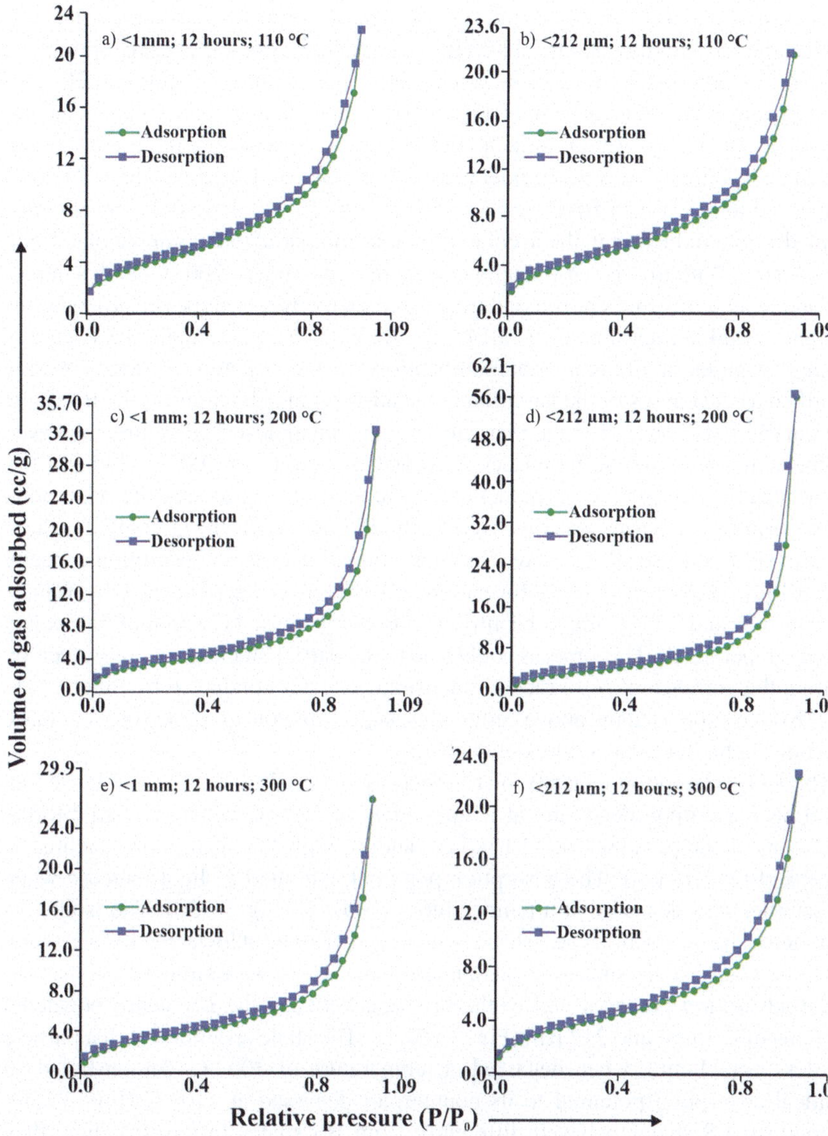

Fig. 3.13 Isotherms of N_2 adsorption and desorption of LS shale crushed to less than 1 mm and less than 212 μm in size and degassed at varying temperatures. Reprinted (adapted) with permission from Singh et al. (2021b); Energy Fuels 2021, 35, 19, 15628–15641. Copyright © 2021 American Chemical Society

Grain size had a remarkably distinct effect on the adsorption–desorption behaviour of N_2 in the Oil-window mature shale (BS shale) (Fig. 3.14a–f). The adsorption and desorption isotherms showed inconsistent behaviour for the < 1 mm particle split that was degassed at temperature of 110 °C, with an open desorption branch and not converging of the adsorption branch (Fig. 3.14a). Conversely, a more constrained pattern of hysteresis loop was evident for < 212 μm sized sample. The < 212 μm sample exhibited rise in BET specific surface area (SSA) and BJH pore volume, coupled with a reduction in the average pore size compared to the sample of < 1 mm particle size. With the rise of the temperature of degassing to 200 °C, both samples exhibited a structure microporous nature, evidenced by steeper slopes in the isotherms at higher relative pressures (Fig. 3.14c, d). As with the LS sample, the excretion of trapped gases or fluids and/or modifications to the structure of organics were identified as the causes of the variation in the adsorption behaviour for the BS shale. Particularly noteworthy was the markedly high residual adsorbed volume observed for the samples of size < 1 mm degassed at temperature of 200 °C (Fig. 3.14d). Upon degassing at 300 °C, a reduction in the separation between the adsorption and desorption isotherms was observed for both crush sizes. Intriguingly, for the < 212 μm split, at lower P/P_0, there was an additional decline in adsorption capacity (Fig. 3.14e). The consistent transformation observed from degassing at temperatures between 110 and 300 °C might be attributed to the increased creation of bitumen at higher temperatures. This process could lead to the partial blockage of pores, thereby limiting the entrance of N_2. Fractal dimensions analysis conducted by Singh et al., (2021b) also supported the above deductions, suggesting the existence of more basic structures at higher temperatures of degassing.

The CO_2 adsorption capacity of LS shales powdered to < 212 μm showed an initial decrease upon degassing at temperatures of 200 °C when compared to the conditions at temperatures of 110 °C, coinciding with the adsorption potential at its peak (Fig. 3.15a–d). The adsorption potential exhibited a slight increase when the samples was degassed at a temperature of 300 °C (Fig. 3.15b). The isotherms exhibited a characteristic Type-I shape, with a steeper slope at lower relative pressure, which suggests a high concentration of micropores. There were similarities observed in the adsorption potential and isotherm shape between the LS shales powdered fractions of < 1 mm and 212 μm (Fig. 3.15a, b). The shale specimen with a particle size less than 1 mm, when degassed at temperature of 300 °C, demonstrated an identical adsorption potential to its counterpart degassed at 110 °C (Fig. 3.15b). In contrast, LS shales behaved differently from BS shales in terms of how they responded to temperature and degassing time (Fig. 3.15c, d). The adsorption capacity of BS shales of particle size of < 212 μm at 110 °C is marginally lower than that of LS shales (3.16 cc/g versus 3.65 cc/g for LS) (Fig. 3.15c). The response to thermal treatment concerning capacity of adsorption of BS shales mirrors that of LS shales, firstly indicating a decreasing trend followed by a raise with rising temperature of degassing. Additionally, the slope of the isotherm decreases with rising temperature, indicating a decrease in smaller micropores (Fig. 3.15a–d). At 200 °C degassing

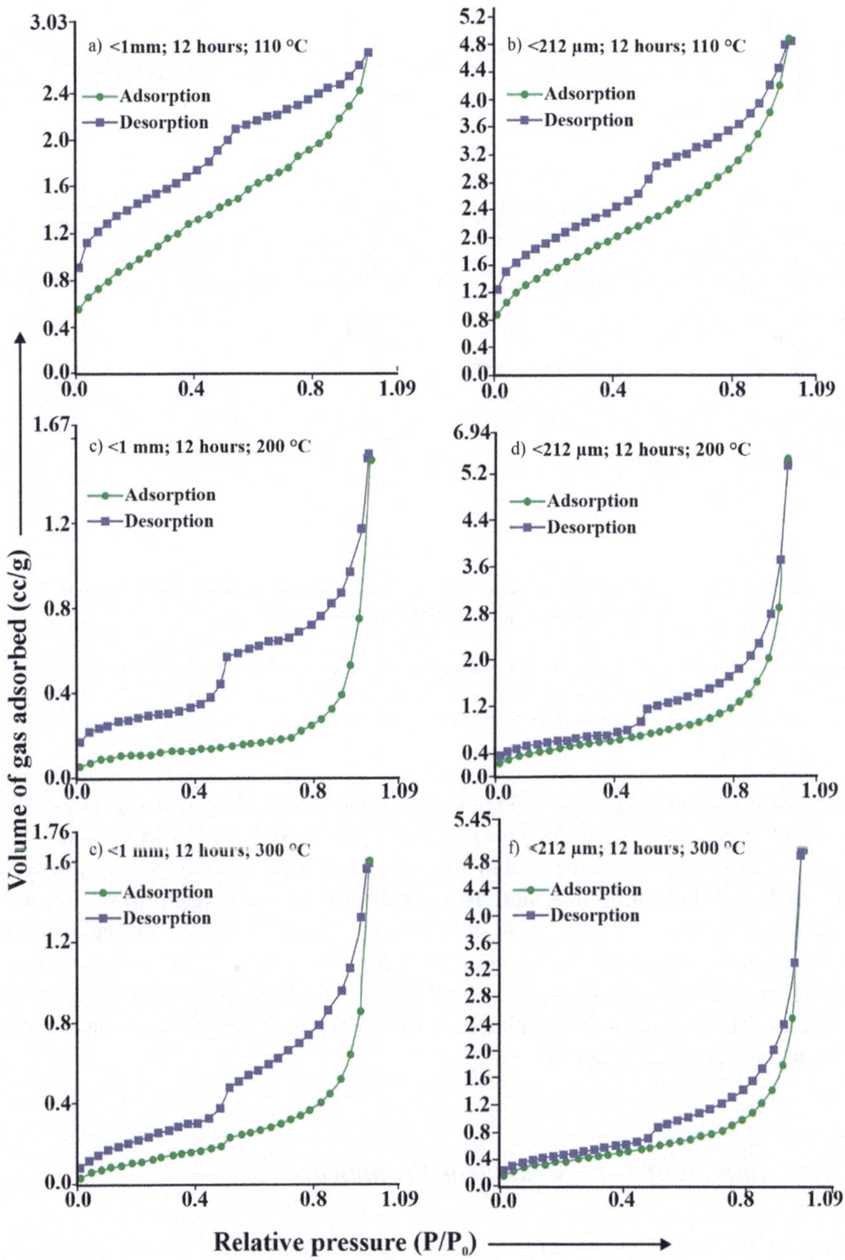

Fig. 3.14 Isotherms of N_2 adsorption and desorption of BS shale crushed to less than 1 mm and less than 212 µm and degassed at varying temperatures. Reprinted (adapted) with permission from Singh et al. (2021b); Energy Fuels 2021, 35, 19, 15628–15641. Copyright © 2021 American Chemical Society

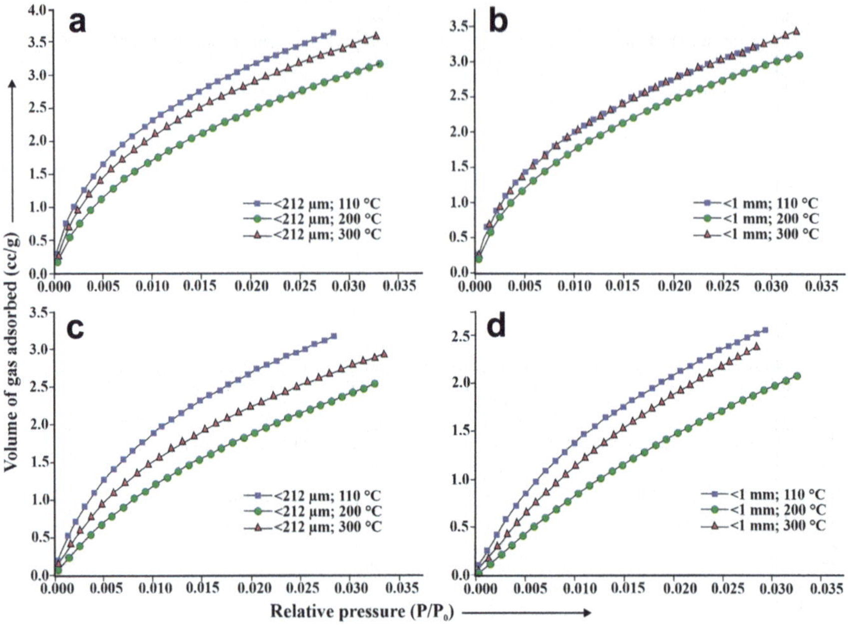

Fig. 3.15 LS (**a**, **b**) and BS (**c**, **d**) shale samples' low-pressure CO_2 gas adsorption isotherms. Reprinted (adapted) with permission from Singh et al. (2021b); Energy Fuels 2021, 35, 19, 15628–15641. Copyright © 2021 American Chemical Society

temperature, the decline in the 0.35 nm peak is partly offset by the emergence of the peak at 0.6 nm, suggesting the enlargement of pores. At 300 °C, there is a notable decrease in the pore volume within the transition zone of micropore−mesopore (0.8−1.5 nm), indicating that there is a potential for these pores to move towards a mesopore system. The BS shale crushed to ~ 1 mm showed a similar thermal response when compared to the samples of crushed-size ~ 212 μm and also nearly evenly sloped isotherm, indicating absence of smaller micropores. In comparison with the < 212 μm crush-sized shales, the adsorption potential is likewise consistently diminished across all temperatures.

3.4.3 Impact of Pore Structural Framework

The pore structure of unconventional shale reservoirs significantly impacts petroleum storage and transport. Inherent surface roughness, as reflected in fractal dimensions and distinctive pore size distributions, are key determinants shaping fluid flow and resource volume characteristics in shale formations. In fractal geometry, the term "D value" typically refers to the fractal dimension. Fractal dimension can be defined as measure of the complexity, irregularity, or self-similarity of a geometric pattern at different scales. In the context of analysing pore structures in materials like shale, the fractal dimension (D value) provides insights into the intricate and convoluted nature of the pore geometry. A higher fractal dimension implies a more intricate and irregular structure, often associated with enhanced surface roughness and intricate patterns within the material. The fractal dimensions derived from the Frenkel-Halsey-Hill (FHH) method find extensive application in assessing the pore structural complexities of shales. Specifically, D1 and D2 fractal dimensions obtained through the FHH method have been widely employed for the computation of surface area and characterization of porous materials (Sakhaee-Pour and Li 2016; Hazra et al. 2019a, b; Li et al. 2019; Zhang et al. 2020; Wood 2021). Nevertheless, numerous studies have highlighted the possibility of underestimating pore fractal dimensions when utilizing the FHH method with N2 adsorption isotherms. This introduces uncertainty into the interpretation of pore texture in specific porous media (Groen et al. 2003; Xiong et al. 2016). The influence of the restricted access of N_2 gas to micropores on fractal dimension calculations remains uncertain.

Singh et al. (2021a) conducted a study to assess the relative advantages of the Wang–Li, Neimark, and FHH fractal determination methods. They employed these methods to analyse organic-rich shales from the Barren Measures Formation. The shale samples from the North Karanpura (NK) and the Raniganj (RN) sub-basins were subjected to N_2 low-pressure gas adsorption (N_2-LPGA) tests. The fractal dimensions obtained through each method were then compared to the pore size distribution, total organic carbon content (TOC), and thermal maturity of the shale samples. Their findings indicated that, across all shale samples, the fractal dimensions determined from the FHH and Wang–Li methods, based on N_2-LPGA isotherm data, exhibited close agreement. In contrast, those measured from Neimark method were deemed unreasonably high, as many values exceeded 3. Consequently, the Neimark method was considered unsuitable for determining fractal dimensions for the shales analyzed in their study. They observed that FHH values were higher than the values obtained using Wang–Li method for the entire adsorption isotherm (D) and the upper segment of the adsorption isotherm (D_2). However, FHH values were lower than the values from Wang–Li method for the lower P/P_0 segment of the adsorption isotherm (D_1). These subtle variations were attributed to fitting errors inherent in the two methods. Consequently, adopting the arithmetic average values of Wang–Li and FHH was deemed reasonable for comparisons with different shale pore properties. Furthermore, they noted that the disparities between the FHH and Wang–Li-calculated D_1 fractal dimensions were slightly more pronounced than those for D and D_2 fractal dimensions. This discrepancy was attributed to the heightened impact of the limited number of isotherm points at the lowest pressures, which tended to decrease D_1 values measured from FHH method and marginally increase D_1 values measured from Wang–Li method. Strong correlations were noted between D, D1, and D2 and T_{max} for the RN shales, whereas only weak correlations were observed for the North NK shales. Across all shale samples, clear correlations were identified between fractal dimensions and specific surface area (SSA), average pore radius, and pore volume.. The higher fractal dimensions observed in the more thermally mature RN shale samples signalled more intricate pore size distributions compared to the NK shales.

References

Babatunde KA, Negash BM, Jufar SR, Ahmed TY, Mojid MR (2022) Adsorption of gases on heterogeneous shale surfaces: A review. J Pet Sci Eng 208:109466

Barrett EP, Joyner LG, Halenda PP (1951) The determination of pore volume and area distributions in porous substances. I. Computations from nitrogen isotherms. J Am Chem Soc 73(1):373–380

Behar F, Vandenbroucke M (1987) Chemical modeling of kerogen. Org Geochem (United Kingdom) 11(1)

Brunauer S, Emmett PH, Teller E (1938) Adsorption of gases in multimolecular layers. J Am Chem Soc 60(2):309–319

Busch A, Gensterblum Y (2011) CBM and CO_2-ECBM related sorption processes in coal: a review. Int J Coal Geol 87(2):49–71

Chalmers GR, Bustin RM, Power IM (2012) Characterization of gas shale pore systems by porosimetry, pycnometry, surface area, and field emission scanning electron microscopy/ transmission electron microscopy image analyses: examples from the Barnett, Woodford, Haynesville, Marcellus, and Doig units. AAPG Bull 96(6):1099–1119

Chandra D, Bakshi T, Bahadur J, Hazra B, Vishal V, Kumar S, Sen D, Singh TN (2023) Pore morphology in thermally-treated shales and its implication on CO_2 storage applications: a gas sorption, SEM, and small-angle scattering study. Fuel 331:125877

Chen Y, Wei L, Mastalerz M, Schimmelmann A (2015) The effect of analytical particle size on gas adsorption porosimetry of shale. Int J Coal Geol 138:103–112

Clarkson CR, Bustin RM, Levy JH (1997) Application of the mono/multilayer and adsorption potential theories to coal methane adsorption isotherms at elevated temperature and pressure. Carbon 35(12):1689–1705

Clarkson CR, Solano N, Bustin RM, Bustin AMM, Chalmers GR, He L, Melnichenko YB, Radliński AP, Blach TP (2013) Pore structure characterization of North American shale gas reservoirs using USANS/SANS, gas adsorption, and mercury intrusion. Fuel 103:606–616

Dąbrowski A (2001) Adsorption—from theory to practice. Adv Coll Interface Sci 93(1–3):135–224

Day SJ, Duffy GJ, Saghafi A, Sakurovs R (2005) Factors controlling CO_2 absorption in Australian coals. In: Greenhouse gas control technologies 7:2221–2224. Elsevier Science Ltd.

de Boer JH (1958) The structure and properties of porous materials. In: Everett DH, Stone FS (eds) Proceedings of the 10th Colston Symposium. Butterworths, London, UK

Dubinin MM (1947) The equation of the characteristic curve of activated charcoal. Dokl Akad Nauk SSSR 55:327–329

Dubinin MM, Astakhov VA (1971) Description of adsorption equilibria of vapors on zeolites over wide ranges of temperature and pressure

Everett DH (1972) Manual of symbols and terminology for physicochemical quantities and units, appendix II: definitions, terminology and symbols in colloid and surface chemistry. Pure Appl Chem 31(4):577–638

Fianu J, Gholinezhad J, Hassan M (2018) Comparison of temperature-dependent gas adsorption models and their application to shale gas reservoirs. Energy Fuels 32(4):4763–4771

Flores RM (2014) Chapter 4—Coalification, Gasification, and Gas Storage. In: Flores RM (Ed) Coal and Coalbed Gas, Elsevier: Boston, MA, USA, pp 167–233

Gregg SJ, Sing KSW, Salzberg HW (1967) Adsorption surface area and porosity. J Electrochem Soc 114(11):279Ca

Groen JC, Peffer LA, Pérez-Ramírez J (2003) Pore size determination in modified micro-and mesoporous materials. Pitfalls and limitations in gas adsorption data analysis. Microp Mesop Mat 60(1–3):1–17

Han H, Cao Y, Chen SJ, Lu JG, Huang CX, Zhu HH, Zhan P, Gao Y (2016) Influence of particle size on gas-adsorption experiments of shales: an example from a Longmaxi Shale sample from the Sichuan Basin, China. Fuel 186:750–757

Harpalani S, Schraufnagel RA (1990) Shrinkage of coal matrix with release of gas and its impact on permeability of coal. Fuel 69(5):551–556

Hazra B, Wood DA, Vishal V, Varma AK, Sakha D, Singh AK (2018a) Porosity controls and fractal disposition of organic-rich Permian shales using low-pressure adsorption techniques. Fuel 220:837–848

Hazra B, Wood DA, Vishal V, Singh AK (2018b) Pore characteristics of distinct thermally mature shales: influence of particle size on low-pressure CO_2 and N_2 adsorption. Energy Fuels 32(8):8175–8186

Hazra B, Chandra D, Singh AK, Varma AK, Mani D, Singh PK, Boral P, Buragohain J (2019a) Comparative pore structural attributes and fractal dimensions of Lower Permian organic-matter-bearing sediments of two Indian basins: inferences from nitrogen gas adsorption. Energy Sour, Part A: Recov, utilization Environ Eff 41(24):2975–2988

Hazra B, Singh AK, Singh PK, Mani D, Boral P, Das M (2019b) Hydrocarbon-generation potential and thermal-maturity of few Indian coals: inferences from organo-petrography and Rock-Eval. Energy Sour, Part A: Recov, Utili, Environ Eff, 1–10

Hazra B, Wood DA, Singh PK, Singh AK, Kumar OP, Raghuvanshi G, Singh DP, Chakraborty P, Rao PS, Mahanta K, Sahu G (2020a) Source rock properties and pore structural framework of the gas-prone Lower Permian shales in the Jharia basin, India. Arab J Geosci 13:1–18

Hazra B, Vishal V, Singh DP (2020b) Applicability of low-pressure CO_2 and N_2 adsorption in determining pore attributes of organic-rich shales and coals. Energy Fuels 35(1):456–464

Hazra B, Vishal V, Sethi C, Chandra D (2022) Impact of supercritical CO_2 on shale reservoirs and its implication for CO_2 sequestration. Energy Fuels 36(17):9882–9903

Heller R, Zoback M (2014) Adsorption of methane and carbon dioxide on gas shale and pure mineral samples. J Unconvent Oil Gas Res 8:14–24

Hiemenz PC, Rajagopalan R (eds) (2016) Principles of colloid and surface chemistry, revised and expanded. CRC Press

Hill DG, Nelson CR (2000) Gas productive fractured shales: an overview and update. Gas Tips 6(3):4–13

Holmes R, Rupp EC, Vishal V, Wilcox J (2017) Selection of shale preparation protocol and outgas procedures for applications in low-pressure analysis. Energy Fuels 31(9):9043–9051

Hu J, Tang S, Zhang S (2016) Investigation of pore structure and fractal characteristics of the Lower Silurian Longmaxi shales in western Hunan and Hubei Provinces in China. J Natural Gas Sci Eng 28:522–535

Hutson ND, Yang RT (1997) Theoretical basis for the Dubinin-Radushkevitch (DR) adsorption isotherm equation. Adsorption 3:189–195

Kang SM, Fathi E, Ambrose RJ, Akkutlu IY, Sigal RF (2011) Carbon dioxide storage capacity of organic-rich shales. SPE J 16(04):842–855

Kapoor A, Ritter JA, Yang RT (1990) An extended Langmuir model for adsorption of gas mixtures on heterogeneous surfaces. Langmuir 6(3):660–664

Kuila U, Prasad M (2013) Specific surface area and pore-size distribution in clays and shales. Geophys Prospect 61(2-Rock Physics for Reservoir Exploration, Characterisation and Monitoring):341–362

Labani MM, Rezaee R, Saeedi A, Al Hinai A (2013) Evaluation of pore size spectrum of gas shale reservoirs using low pressure nitrogen adsorption, gas expansion and mercury porosimetry: a case study from the Perth and Canning Basins, Western Australia. J Petrol Sci Eng 112:7–16

Li Q, Pang X, Tang L, Chen G, Shao X, Jia N (2018) Occurrence features and gas content analysis of marine and continental shales: a comparative study of Longmaxi Formation and Yanchang Formation. J Nat Gas Sci Eng 56:504–522
Li Y, Wang Z, Pan Z, Niu X, Yu Y, Meng S (2019) Pore structure and its fractal dimensions of transitional shale: a cross-section from east margin of the Ordos Basin, China. Fuel 241:417–431
Liu X, Zhang D (2019) A review of phase behavior simulation of hydrocarbons in confined space: implications for shale oil and shale gas. J Nat Gas Sci Eng 68:102901
Liu A, Fu X, Wang K, An H, Wang G (2013) Investigation of coalbed methane potential in low-rank coal reservoirs–Free and soluble gas contents. Fuel 112:14–22
Lowell S, Shields JE, Thomas MA, Thommes M (2006) Characterization of porous solids and powders: surface area, pore size and density, Vol 16. Springer Science & Business Media
Lu XC, Li FC, Watson AT (1995) Adsorption measurements in Devonian shales. Fuel 74(4):599–603
Luffel DL, Guidry FK (1992) New core analysis methods for measuring reservoir rock properties of Devonian shale. J Petrol Technol 44(11):1184–1190
Mastalerz M, He L, Melnichenko YB, Rupp JA (2012) Porosity of coal and shale: insights from gas adsorption and SANS/USANS techniques. Energy Fuels 26(8):5109–5120
Mastalerz M, Hampton L, Drobniak A, Loope H (2017) Significance of analytical particle size in low-pressure N_2 and CO_2 adsorption of coal and shale. Int J Coal Geol 178:122–131
McCarty RD, Arp VD (1990) A new wide range equation of state for helium. Advances in cryogenic engineering: Part A & B. Springer, US, Boston, MA, pp 1465–1475
Michels A, Wouters H (1941) Isotherms of helium between 0 and 150 C up to 200 Amagat. Physica 8(8):923–932
Moore TA (2012) Coalbed methane: a review. Int J Coal Geol 101:36–81
Mosher K, He J, Liu Y, Rupp E, Wilcox J (2013) Molecular simulation of methane adsorption in micro-and mesoporous carbons with applications to coal and gas shale systems. Int J Coal Geol 109:36–44
Mudoi MP, Sharma P, Khichi AS (2022) A review of gas adsorption on shale and the influencing factors of CH_4 and CO_2 adsorption. J Petrol Sci Eng 217:110897
Neimark AV, Ravikovitch PI, Vishnyakov A (2000) Adsorption hysteresis in nanopores. Phys Rev E 62(2):R1493
Odoh SO, Cramer CJ, Truhlar DG, Gagliardi L (2015) Quantum-chemical characterization of the properties and reactivities of metal–organic frameworks. Chem Rev 115(12):6051–6111
Ozdemir E (2004) Chemistry of the adsorption of carbon dioxide by Argonne Premium coals and a model to simulate carbon dioxide sequestration in coal seams. Doctoral dissertation, University of Pittsburgh
Pan Z, Connell LD (2015) Reservoir simulation of free and adsorbed gas production from shale. J Nat Gas Sci Eng 22:359–370
Perera MSA, Ranjith PG, Choi SK, Airey D, Weniger P (2012) Estimation of gas adsorption capacity in coal: a review and an analytical study. Int J Coal Prep Util 32(1):25–55
Pirngruber G (2016) Physisorption and pore size analysis. Characterization of Porous Solids-Characterization of catalysts and surfaces. Institut Francais du Petrole, 68 p
Polanyi M (1932) Section III. Theories of the adsorption of gases. a general survey and some additional remarks. Introductory paper to section III. Trans Faraday Soc 28:316–333
Ravikovitch PI, Domhnaill SÓ, Neimark AV, Schüth F, Unger KK (1995) Capillary hysteresis in nanopores: theoretical and experimental studies of nitrogen adsorption on MCM-41. Langmuir 11(12):4765–4772
Ross DJ, Bustin RM (2009) The importance of shale composition and pore structure upon gas storage potential of shale gas reservoirs. Mar Pet Geol 26(6):916–927
Rouquerol J, Rouquerol F, Llewellyn P, Maurin G, Sing K (2013) Adsorption by powders and porous solids: principles, methodology and applications. Academic Press
Ruthven DM (2006) Fundamentals of adsorption equilibrium and kinetics in microporous solids. In: Adsorption and diffusion. Springer, Berlin, Heidelberg, pp 1–43
Sakhaee-Pour A, Li W (2016) Fractal dimensions of shale. J Nat Gas Sci Eng 30:578–582

Schettler PD, Parmoly CR (1990) The measurement of gas desorption isotherms for Devonian shale. GRI Devonian Gas Shale Tech Rev 7(1):4–9

Schmitt M, Fernandes CP, da Cunha Neto JA, Wolf FG, dos Santos VS (2013) Characterization of pore systems in seal rocks using nitrogen gas adsorption combined with mercury injection capillary pressure techniques. Mar Pet Geol 39(1):138–149

Sing K (2001) The use of nitrogen adsorption for the characterisation of porous materials. Colloids Surf, A 187:3–9

Sing KS (1985) Reporting physisorption data for gas/solid systems with special reference to the determination of surface area and porosity (recommendations 1984). Pure Appl Chem 57(4):603–619

Sing KS (2004) Characterization of porous materials: past, present and future. Colloids Surf, A 241(1–3):3–7

Singh DP, Chandra D, Vishal V, Hazra B, Sarkar P (2021a) Impact of degassing time and temperature on the estimation of pore attributes in shale. Energy Fuels 35(19):15628–15641

Singh DP, Wood DA, Hazra B, Singh PK (2021b) Pore properties in organic-rich shales derived using multiple fractal determination models applied to two Indian Permian basins. Energy Fuels 35(18):14618–14633

Song W, Yao J, Ma J, Li A, Li Y, Sun H, Zhang L (2018) Grand canonical Monte Carlo simulations of pore structure influence on methane adsorption in micro-porous carbons with applications to coal and shale systems. Fuel 215:196–203

Srodon J, McCarty DK (2008) Surface area and layer charge of smectite from CEC and EGME/H2O-retention measurements. Clays Clay Miner 56(2):155–174

Thommes M, Smarsly B, Groenewolt M, Ravikovitch PI, Neimark AV (2006) Adsorption hysteresis of nitrogen and argon in pore networks and characterization of novel micro-and mesoporous silicas. Langmuir 22(2):756–764

Vishal V, Chandra D, Bahadur J, Sen D, Hazra B, Mahanta B, Mani D (2019) Interpreting pore dimensions in gas shales using a combination of SEM imaging, small-angle neutron scattering, and low-pressure gas adsorption. Energy Fuels 33(6):4835–4848

Vishnyakov A, Neimark AV (2003) Nucleation of liquid bridges and bubbles in nanoscale capillaries. J Chem Phys 119(18):9755–9764

Wang Y, Zhu Y, Liu S, Zhang R (2016) Pore characterization and its impact on methane adsorption capacity for organic-rich marine shales. Fuel 181:227–237

Wang H, Qu Z, Yin Y, Bai J, Yu B (2019) Review of molecular simulation method for gas adsorption/desorption and diffusion in shale matrix. J Therm Sci 28:1–16

Wei M, Xiong Y, Zhang L, Li J (2016) The effect of sample particle size on the determination of pore structure parameters in shales. Int J Coal Geol 163:177–185

Wood DA (2021) Techniques used to calculate shale fractal dimensions involve uncertainties and imprecisions that require more careful consideration. Adv Geo-Energy Res 5(2):153–165

Wu K, Chen Z, Li X (2015) Real gas transport through nanopores of varying cross-section type and shape in shale gas reservoirs. Chem Eng J 281:813–825

Xiong Q, Baychev TG, Jivkov AP (2016) Review of pore network modelling of porous media: experimental characterisations, network constructions and applications to reactive transport. J Contam Hydrol 192:101–117

Yang F, Ning Z, Liu H (2014) Fractal characteristics of shales from a shale gas reservoir in the Sichuan Basin, China. Fuel 115:378–384

Yang Y, Liu S (2020) Laboratory study of cryogenic treatment induced pore-scale structural alterations of Illinois coal and their implications on gas sorption and diffusion behaviors. J Petrol Sci Eng 194:107507

Yi J, Bao H, Zheng A, Zhang B, Shu Z, Li J, Wang C (2019) Main factors controlling marine shale gas enrichment and high-yield wells in South China: a case study of the Fuling shale gas field. Mar Pet Geol 103:114–125

Yu W, Sepehrnoori K, Patzek TW (2016) Modeling gas adsorption in Marcellus shale with Langmuir and BET isotherms. SPE J 21(02):589–600

Yu H, Xu H, Fan J, Zhu YB, Wang F, Wu H (2020) Transport of shale gas in microporous/nanoporous media: molecular to pore-scale simulations. Energy Fuels 35(2):911–943

Yuan J, Luo D, Feng L (2015) A review of the technical and economic evaluation techniques for shale gas development. Appl Energy 148:49–65

Zhang T, Ellis GS, Ruppel SC, Milliken K, Yang R (2012) Effect of organic-matter type and thermal maturity on methane adsorption in shale-gas systems. Org Geochem 47:120–131

Zhang J, Tang Y, He D, Sun P, Zou X (2020) Full-scale nanopore system and fractal characteristics of clay-rich lacustrine shale combining FE-SEM, nano-CT, gas adsorption and mercury intrusion porosimetry. Appl Clay Sci 196:105758

Zhou F, Xia T, Wang X, Zhang Y, Sun Y, Liu J (2016) Recent developments in coal mine methane extraction and utilization in China: a review. J Nat Gas Sci Eng 31:437–458

Chapter 4
Imaging and Visualization

Unconventional reservoirs like gas shale and coal are commonly classified as fine grained sedimentary rocks. Gases in these reservoirs are stored within the pores hosted in organic matter or mineral components. Because of the extremely small grain size, the pores in these reservoirs are also extremely small (in the order of nanometres) compared to conventional reservoirs. While the pores in these reservoirs serve as storage space for methane, the natural and induced fractures serve as primary conduits for migration of the stored gases. These fractures are usually larger in scale and connects the pores in the matrix to facilitate fluid flow. Imaging techniques not only help us visualize the pores and other features in shale, but also helps us quantify their volume, surface area, interconnectivity in 3D and 2D. Given the extremely small pore sizes in coal and shale, specialized imaging techniques are leveraged to probe into such fine features. The evolution of microscopy techniques and associated image processing workflow over the last few decades have enabled the scientific community to not only visualize static images, but also dynamic changes in chemical, thermal and flow attributes of shale and coal. Applicability of ML techniques have further helped merge imaging data acquired across multiple scales to construct a better understanding of these reservoir rocks. Chandra and Vishal (2021) presented a summary of different imaging techniques and their applicability to scan different features in unconventional reservoir rocks like coal and shale (Fig. 4.1).

B. Hazra et al., *Unconventional Hydrocarbon Reservoirs: Coal and Shale*, Petroleum Engineering, https://doi.org/10.1007/978-3-031-53484-3_4

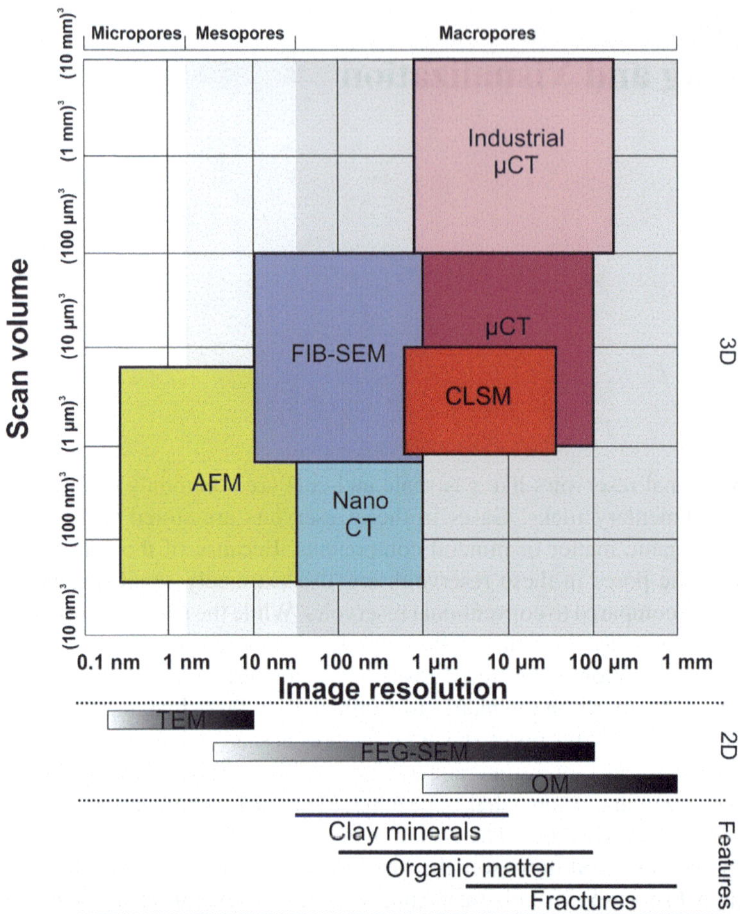

Fig. 4.1 Common 2D and 3D imaging techniques used for studying unconventional reservoir rocks with their scanning range and resolution. Reprinted from Earth-Science Reviews, Vol. 217, Chandra and Vishal (2021), A critical review on pore to continuum scale imaging techniques for enhanced shale gas recovery, Page No. 103638, Copyright (2021), with permission from Elsevier

Compared to 3D imaging techniques, 2D imaging can generally go for higher resolution, which is useful for imaging nanoscale pores, albeit with a trade-off for the depth information of the feature. However, for imaging small pores, often such trade-off is acceptable. One such trade-off across all imaging techniques is the compromise of imaging resolution while going for larger field of view (FOV). Given the heterogeneity in shale and coal, opting for a larger FOV can often bring the selected scan area closer to the representative elementary volume (REV). However, this approach is not suitable for imaging finer features due to the inverse relationship between FOV and imaging resolution. For example, a lower FOV can help image nanopores with a higher detail, but will fail to show the neighbouring grains or the host mineral/organic matter. Current technologies like mosaic imaging can seamlessly stitch numerous individual scan areas, which can counter the trade-off between image resolution and FOV, however such process is expensive, time consuming, and produces huge volume of data, which is computationally expensive.

4.1 Sample Preparation for Imaging

Similar to any other experimental technique, different imaging techniques also require dedicated size and shape of sample, which is useful to keep in mind while making the judgement of using certain imaging techniques. As mentioned before, shale and coal are highly heterogeneous, hence using different parts of the sample for different imaging approaches may not give similar inferences. A convenient workflow proposed by Chandra and Vishal (2021) illustrated how to efficiently use a specimen for preparing different samples required for multiple 2D and 3D imaging techniques (Fig. 4.2) and different scale dependent information that can be extracted from each of these techniques.

Their workflow indicates that preparing samples for 3D imaging is much easier compared to 2D imaging. Since 2D imaging techniques can usually go to higher resolution, they are very sensitive to surface roughness, impurities and surface conductivity (for imaging with charged particles e.g. SEM, TEM etc.). For most imaging techniques, samples are trimmed to a regular dimension with cutting blades, however the size of the specimen depends on the imaging process to be adopted. 3D imaging techniques (e.g. X ray/PET CT) are non destructive in this aspect, except focussed ion beam scanning electron microscopy (FIB-SEM), meaning that any regular or irregular sized specimen can be placed in the samples holder and the user can either scan the entire sample or zoom into the desired part of the sample to scan in a higher resolution. However, a major limitation lies in the source to sample distance, where more magnification can be achieved by bringing the source closer to the sample, but the dimension of the sample itself can pose an obstacle in such case. Among 3D imaging, FIB-SEM is often included, however the working principle of FIB-SEM is different, where thin slices of the sample surface is physically eroded by powerful ion guns to reveal a new surface. These consecutive surfaces are imaged in a similar way as standard SEM imaging and then stacked together to construct a 3D volume. 2D imaging (e.g. optical microscopy, SEM/TEM imaging) requires special sample preparation protocol, where thin slices of samples are essential, which makes it a destructive process. The usual sample thickness are ~ 20 μm for glass-mounted thin sections, ~ 1 mm for samples used in SEM imaging, and ~ 10–20 nm thin wafer for TEM imaging. Moreover for SEM and TEM imaging, the specimen are sputter coated with conductive metals (Au, Ag, Pt etc.) for even conductivity throughout the surface, reducing effects of local charging. We hereby present a brief description of sample preparation specifics for commonly used imaging techniques:

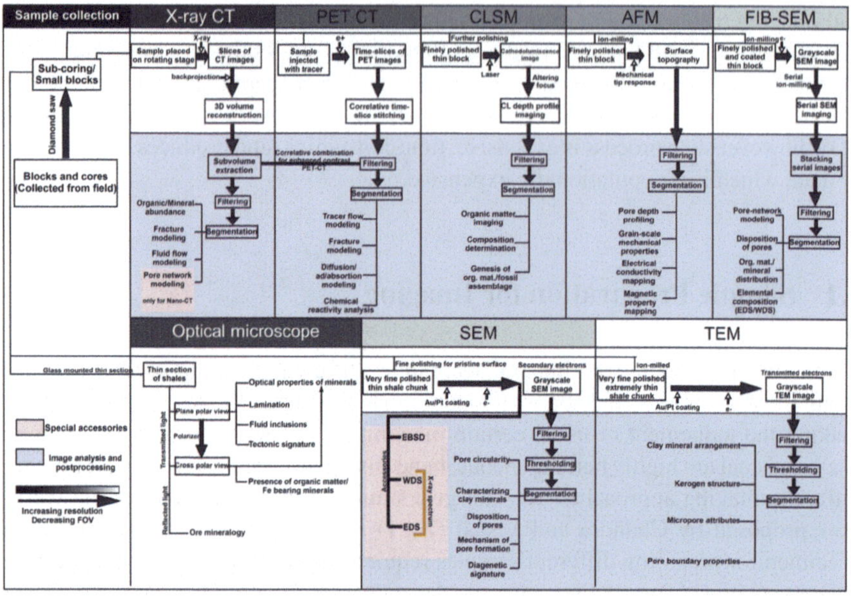

Fig. 4.2 A brief workflow of sample preparation for most used 2D and 3D imaging techniques for shale and coal, and information extracted from each of these techniques. Reprinted from Earth-Science Reviews, Vol. 217, Chandra and Vishal (2021), A critical review on pore to continuum scale imaging techniques for enhanced shale gas recovery, Page No. 103638, Copyright (2021), with permission from Elsevier

a. **X-ray CT imaging:** Resolution of a micro X-ray CT can go as small as 1 μm, hence the sample size should be considerably larger than the highest resolution. Nano CTs can go for even higher resolution (~ 100 nm) and can accommodate even smaller samples. Due to limitation in resolution, pores in coal and shales cannot be imaged with X-ray CT, however, quantification of cracks and core scale heterogeneity and anisotropy can be easily quantified with such technique. Usual sample size scanned with X-ray CT is standard cylindrical cores/coreplugs which can range between a diameter of 25–60 mm and a length of 50–120 mm. In modern CT scanners, samples up to 1 m length can also be scanned, where the instrument images separate sections of the sample and stitches them together during reconstruction. During imaging, the sample rotates or the source-detector assembly rotates around the sample. In either case, having a cylindrical shape helps deciding the source-sample distance easily without the risk of the sample touching the source. Also, a curved surface creates lesser beam hardening, which is more prone to form in sharper edges.
b. **PET imaging**: Sample dimension for positron emission tomography (PET) follows requirements for X-ray CT. However, PET requires presence of a radionuclide within the scanned object, which can emit positron during imaging, which are recorded with the photomultiplier tubes arranged in a circular orientation.

Since the components inherently present in shale or coal doesn't emit positrons spontaneously, application of PET is only limited for flow imaging, where a tracer radionuclide flows through a crack and the instrument images the pathway of the fluid over time.

c. **Confocal microscopy**: Samples size for confocal microscopy is usually similar to the size requirement of optical microscopy. Since the sample holder is open from all sides, it can accommodate specimen of a few centimetres in length and width. The depth of the specimen is usually limited to a few microns.
d. **Atomic force microscopy**: Atomic force microscopy (AFM) is a versatile tool for microscale surface profiling, mechanical property characterization as well as electrical/magnetic profiling. AFM works with a tip and cantilever setup, which hovers on the sample surface or stays in contact with the surface depending on the requirement. Analysis with AFM is extremely surface sensitive, any major surface roughness can cause error in results and sometime may break the tip. Samples mounted in AFM can be of a similar dimension as of optical microscopy and can be a little bit thicker (up to a few mm). Surface roughness should be lesser than 5 nm, which requires extensive fine polishing before mounting the specimen in AFM.
e. **Scanning electron microscopy**: Scanning electron microscopy (SEM) is a popular tool used for high resolution imaging in different materials. SEM imaging requires to evacuate the electron column and hence the sample size used is very small compared to the size requirements for optical microscopy. Usual sample size is ~ 1 cm × 1 cm with ~ 2 mm thickness. The samples are extensively polished to minimize surface roughness and are also sputter coated with conductive metals for even charge distribution on the surface.
f. **Transmission electron microscopy**: Transmission electron microscopy (TEM) works in similar principle as SEM, however the resolution is even higher, often in terms of a few nanometres (0.2–0.5 nm). The sample prepared for TEM therefore needs to be of even smaller size compared to the one prepared for SEM. Preparing such wafer thin and small sample of fissile material like shales and coal are often troublesome, which restricts use of TEM for these applications. Sample preparation often involves ion milling, polishing and sputter coating at high vacuum, which is expensive.
g. **Focussed ion beam SEM**: Focussed ion beam SEM or FIB-SEM is a novel technique which utilizes the combination of simultaneous ion milling and SEM imaging to provide 3D information. This is a destructive process where small samples of shale/coal (smaller than the ones used in SEM) are loaded in an SEM connected with an ion milling setup often using charged Ar or Ga. Surface imaging similar to conventional SEM is performed and the imaged surface is eroded to image the surface beneath. Using a sample with larger surface area often takes longer to mill out, and organic matter within the samples often show burning artifact due to the temperature generated during ion milling. Although there are ways to remediate this issue (e.g. placing the sample within a freezing bath), such facilities are limited and expensive.

4.2 Use of 2D Imaging in Coal/Shale Characterization

The most common form of 2D imaging of geomaterials is optical microscopy of glass-mounted thin sections of rocks. Although this method doesn't suit very high magnification like SEM, we can obtain key information on the organic and mineral matrix distribution, grain sizes, bedding orientations and in some cases also identify mineral compositions (for medium/course grained rocks). This technique is widely used for different type of rocks and ore minerals to determine their genesis, geochemical alteration, or mineral's optical properties (relief, birefringence, twinning, cleavage, etc.) As the light source used for optical microscopy is in the visible spectra (~ 390–700 nm wavelength), the maximum resolution is theoretically limited to 0.23 μm (Bultreys et al. 2016). SEM is a widely used versatile tool for high resolution imaging of geomaterials. Organic-rich shale or coal samples pose challenge with optical microscopy, as the organic matter appears as dark opaque material in visible light spectra. Polishing such weak rocks while preparing thin section pose challenges, where the organic matter might peel off from the glass mount. Most commonly for ultrafine-grained rocks like coal and shale, optical microscopy is coupled with other techniques to obtain comprehensive idea about heterogeneity, anisotropy and depositional environment (Fig. 4.3) of the rocks (Buchardt and Lewan 1990; Chandra et al. 2020b; Loucks and Ruppel 2007). Thin sections also acts as reconnaissance for selecting areas for SEM/TEM sample preparation.

SEM is a versatile high resolution imaging technique, which is extensively used in geomaterials. In case of shale and coal, specific use of SEM lies in detecting pores in the organic/mineral matrix. SEM provides grayscale images where brighter pixels indicate conductive/denser phases and darker pixels indicate lighter phases. In general minerals appear brighter as they are denser than organic matter, however, organic matter is denser than pores (Hazra et al. 2023), which in theory is a void filled with air. Therefore in a system with pores, organic matter and minerals, pores will be darkest, followed by organic matter and minerals. SEM has been extensively used to quantify pore connectivity and pore shape anisotropy in coal and shales (Chandra et al. 2020a; Chandra and Vishal 2020, 2022; Davudov et al. 2020; King et al. 2015; Song et al. 2019; Vishal et al. 2019; Wu et al. 2020; Zou et al. 2020). It has also been extensively used to characterize changes in pore structure, mineral crystallization or alteration before and after chemical or thermal treatments (Chandra et al. 2023, 2021; Cheng et al. 2021; Dai et al. 2020; Fatah et al. 2022; Li et al. 2023; Yin et al. 2016) Another advantage of using SEM is its customizability. SEM can be coupled with wavelength-dispersive spectroscopy (WDS) or energy-dispersive spectroscopy (EDS), which assists in element mapping of the visible surface, and can also be automated using QEMSCAN, MLA, MAPS and similar algorithms (Allen et al. 2018; Li et al. 2018a, b, 2022a, b; Lin et al. 2022; Tang et al. 2016). Electron bombardment on the specimen during SEM imaging creates a wide spectrum of electron liberation from the sample surface including Auger electrons, backscattered electrons (BSE), secondary electrons (SE) and cathodoluminescence (CL). Principle of TEM is very

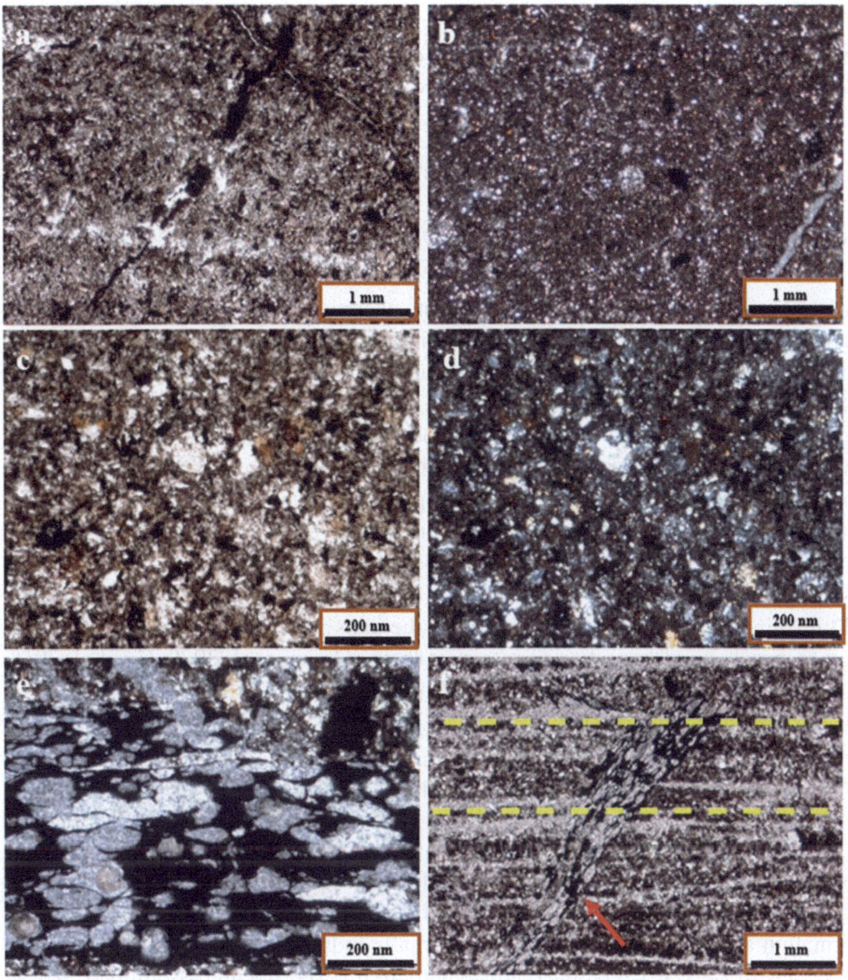

Fig. 4.3 Optical microscopy of shale samples of different grainsize and different organic-mineral disposition. Images are taken under plane polarized light. The dark entities in the figure indicates organic matter. Reprinted (adapted) with permission from Chandra et al. (2020b); Energy Fuels 2020, 34, 7, 8085–8098. Copyright © 2020 American Chemical Society

similar to SEM, however in this case the electrons penetrate through the wafer-thin specimen and produces elastic/inelastic scattering and direct beam, which is used to image the material. Similar to SEM, TEMs are also coupled with additional units like electron energy loss spectroscopy (EELS) to enhance characterization of materials using the diffracted and scattered electrons. Although use of TEM for characterizing coal and shale is limited, it can provide details about pore surface properties and its change thereof in an unprecedented resolution. TEM has been used to characterize inherent chemical heterogeneity and different macerals in coal (Duber

2011; Kwiecińska et al. 2019; Kwiecińska and Petersen 2004; Pusz et al. 2014; Taylor 1966; Taylor et al. 1991, 1998; Taylor and Teichmüller 1993). Geochemical evolution of organic rich shales have been quantified using STEM and TEM (Bernard et al. 2012a, 2013; b).

4.3 Use of 3D Imaging in Coal/Shale Characterization

While transitioning from 2 to 3D imaging techniques, the first technique worth mentioning is FIB-SEM, which operates similar to a SEM with in-situ ion-milling capabilities. Each layer of specimen is scanned in a similar way as SEM, however as consecutive deeper layers are scanned, this technique can provide depth information up to a few mm. This is useful for making 3D volumes of the mesopores in organic matter in shale, however the smaller scan volume restricts the technique to quantify the heterogeneity and anisotropy in shale and coal. FIB-SEM has been extensively used to construct pore network model (PNM) of mesopores in shale and coal, which can be further used to model mass transport through the pores and assess their connectivity. FIB-SEM has often been used in combination with other 3D imaging tools in an attempt to merge the gaps in scale of imaging to obtain a comprehensive understanding of reservoir rocks (Garum et al. 2020; Gu et al. 2015; Kelly et al. 2016; Li et al. 2017, 2020; Liu et al. 2017; Saif et al. 2017; Sun et al. 2020; Xiong et al. 2020; Zhao et al. 2022; Zhou et al. 2016, 2017).

X-ray CT is another common imaging technique for geomaterials to visualize and quantify 3D distribution of heterogeneity, veining, fracture network and pore structure as well as its connectivity. Since shale and coal are nanoporous, imaging pore structure in these rocks are tricky, however some researchers have been able to perform this with extremely high resolution imaging (Arif et al. 2021; Garum et al. 2020; Liu et al. 2023; Tiwari et al. 2013). Dynamic processes like flow, crack formation and propagation as well as thermally induced changes can also be easily imaged with X-ray CT imaging coupled with in situ experimental assembly (Cai et al. 2014; Cała et al. 2016; Duan et al. 2022; Liu et al. 2018; Vishal and Chandra 2022; Wang et al. 2018a, b, 2019; Yu et al. 2019). Transport behavior through fractures in shale and coal can also be imaged with PET, where a radionuclide tracer doped fluid acts as the fluid and fast timelapse image volumes are recorded to track the movement of the tracer (Boutchko et al. 2012; Brattekås et al. 2017; Fernø et al. 2015; Joss and Pini 2018; Kulenkampff et al. 2008; Li et al. 2019; Lippmann-Pipke et al. 2017; Pini and Joss 2019). PET is often better than X-ray CT to track fluid flow, since lab based PET imaging are often faster than CT imaging, which enables uncover many finer details of fluid movement. X-ray imaging might pose difficulty when the fluid has the same density as any mineral or organic matter present within the sample. However, PET has coarser resolution compared to X-ray CT, which leads to adoption of channel fusion algorithm to convolute X-ray and positron based CT to get the best of both worlds.

AFM is an emerging tool for characterizing surface or near surface properties of geomaterials and has been extensively used for metals and synthetic polymers for a long time. AFM imaging can be performed by contact, non-contact and tapping mode. In the contact mode, the tiny repulsion force applied on the cantilever by the sample surface is recorded and a map of the scan area based on the repulsion force is recorded. Depending on the chemical compositional variation or surface unevenness (due to presence of pores), the repulsion force varies significantly. In the non-contact mode, instead of the repulsion force, the attraction force between the sample surface and the cantilever tip is recorded (Javadpour et al. 2012). AFM has been extensively used for characterizing kerogen, minerals, pores and innate fractures in shale and coal by tapping method and contact based method (Abarghani et al. 2020; Kumar et al. 2018; Li et al. 2018a, b; Liu et al. 2019a, b).

4.4 Image Processing Techniques

So far we have discussed about the state of the art imaging techniques in detail, however, recording images only provide qualitative assessments of the studied sample. Image processing plays a huge part in extracting quantitative information about the subject/feature being studied. There are many commercial and non-commercial software available for image processing, stacking and data acquisition. It's also important to mention that the rise in correlative microscopy, merging different imaging techniques across multiple scales have been possible largely due to compatible softwares which can tag locations and register image data seamlessly. Most common image processing softwares include Avizo (https://www.fei.com/software/avizo3d/%C2%A0), Dragonfly (https://www.theobjects.com/dragonfly/index.html), ImageJ (https://imagej.net/), MATLAB Image Processing Toolbox (https://nl.mathworks.com/products/image.html), Gwyddion (http://gwyddion.net/) and so on. Common image processing algorithms used for coal/shale imaging in 2D and 3D are briefly discussed in the following paragraph.

An essential operation used in image processing is image filtering. Filtering are mathematical operations which can be applied by the user to remove artifacts like noises, fringing, chromatic aberration, beam hardening, etc. Often a combination of filters are adopted to enhance the quality of the image and improve the quantification. Low-pass or high-pass filters can be applied to attenuate high frequency and low frequency signals in the image respectively. A median filter can smoothen the features in the image based on illuminance data from the neighbouring pixels, which can remove some of the salt and pepper noise. An unsharp masking filter can be applied followed by a median filter to preserve the edge information which gets lost due to application of median filter.

Feature extraction and quantification from 3D or 2D images is facilitated by the application of segmentation. The user establishes a threshold for pixel values depending on the pixel value range of the features intended to be extracted. For example, extracting pore structure from the mineral and organic matrix requires setting the minimum and maximum value of the pore spaces as a threshold and subtract it from the original image. More complexity arises when there are two or more classes lying within a same pixel value range. In that case ML based techniques are leveraged, where shape based thresholding is performed. For example, pores and open fractures has similar greyscale values, since both are voids filled with air, however, fractures are planar features and pores are ellipsoidal features. Based on this hypothesis, a ML algorithm can be trained to differentiate between pores and fractures. Watershed segmentation is a useful technique which can fine-tune manual thresholding, that may struggle to extract features accurately because of overlapping grains or diffused grain boundaries. In such cases, the watershed algorithm can efficiently distinguish between two or more fused pores, throats connecting multiple pores and grains in contact in porous media.

ML and automated segmentation, is an evolving domain in classifying porous media, which included sedimentary rocks like coal and shale. ML based labelling and segmentation techniques have been found to be more accurate and fast compared to manual segmentation, which contains human error (Asadi and Beckingham 2021; Ganguly et al. 2020; Kazak et al. 2021; Knaup et al. 2019; Wu et al. 2019). ML based techniques are also integrated into standard image processing workflow of different commercial and non-commercial softwares. Further advancement in application of ML in computer vision will make segmentation and quantification even easier.

The strength of imaging lies in successful combination of multiple multiscale experimental or imaging techniques. Image based quantification alone cannot paint a full picture on the reservoir character, but can help with the missing information in standard reservoir characterization. There have been multiple instances where different imaging techniques are combined or one or more imaging techniques are combined with some experimental techniques for a more comprehensive understanding of coal/shale. One such technique is elaborated in the following section.

4.5 Optical-Electron Correlative Microscopy

Optical-electron correlative microscopy is a type of petrographic technique used for evaluating same field of view for identification and characterization of the organic matter present in shales using both light and electron microscopy (Hackley et al. 2017; Valentine and Hackley 2019; Liu et al. 2022; Sethi et al. 2023a). Figure 4.4 shows the correlative image set of a shale sample from Rajmahal Basin, India. Using the correlative technique, the same field of view which was observed and marked during optical microscopy under brightfield reflected light (Fig. 4.4a) and fluorescence light (Fig. 4.4b) is viewed under SEM at 5, 10, and 15 kV acceleration voltages. It can be observed that the dispersed organic matter (DOM) and mineral matter (MM) present in the shale sample are easily discernible at 15 kV in BSE (backscattered electrons) scanning mode (Fig. 4.4c) and at 10 kV in SE2 (secondary electrons) scanning mode (Fig. 4.4f). The organic and mineral phases cannot be distinguished with certainty at 10 kV acceleration voltage in BSE scanning mode (Fig. 4.4E). Decreasing the accelerating voltage to 5 kV, results in further deterioration of contrast between the organic and mineral phases (Fig. 4.4g). On the other hand, the contrast between the organic and mineral phases is poor when SE2 scanning is used at 15 and 5 kV acceleration voltages (Fig. 4.1d, h, respectively). The grayscale contrast between the organic and mineral matter can be explained through post-acquisition image analysis (Fig. 4.5) using ImageJ (version 1.53 k). In Fig. 4.5c, it is evident that the gray-value peaks are at their highest when the line profile (blue line) traverses through mineral phases and at their lowest when passing through organic phases. The mean grayscale value of the organic matter present in the shale sample (Fig. 4.5a) was observed to be 100. Selection of appropriate working distance (WD) during electron microscopy is also important. Sethi et al. (2023b) in their optical-electron correlative study of shales form Rajmahal and Jharia basins of India stressed that depending upon the SEM instrument the appropriate WD should be chosen to achieve reliable distinction between the organic and mineral phases. In their analysis of the shale samples at working distances of 11 mm (Fig. 4.6a, b), 13 mm (Fig. 4.6c, d), and 15 mm (Fig. 4.6e, f), they found that the WD of 13 mm is the most accurate for separating mineral matter from DOM (Fig. 4.6c, d).

Identification of different organic matter types present in the shale/coal samples under SEM alone can be a daunting task to achieve. Although SEM has been widely used in different studies: (a) evaluation of organic matter-hosted porosity (e.g., Loucks et al. 2009; Milner et al. 2010; Milliken et al. 2014; Inan et al. 2018); (b) distinguishing mineral phases (e.g., Camp 2016); (c) micromechanics (e.g., Cheng et al. 2022), etc., only in the recent years correlative microscopy made it possible to identify different organic matter types through SEM (Hackley et al. 2017; Valentine and Hackley 2019; Sethi et al. 2023a, b). Additionally, with the use of the correlative technique, the observed microstructural properties of the organic matter could now be related to the specific macerals.

Figure 4.7 shows the correlative image-sets of a shale sample from Rajmahal Basin of India. Through the correlative technique, vitrinite present in the shale sample were first identified and marked as region of interest (ROI) (Fig. 4.7a: ROI-1, b: ROI-1) under an optical microscope and then observed under the SEM (Fig. 4.7c, d). The marked vitrinite fragment under SEM appears dark in BSE scanning mode and exhibits negative relief in SE2 scanning mode in contrast to the surrounding mineral phases (Fig. 4.7c, d, respectively). Under SEM, EDS analysis indicated the presence of clay mineral within the vitrinite maceral (Fig. 4.7c: EDS Spot 1). However, without prior marking under optical microscope using the correlative technique, the vitrinite maceral would have been mistakenly identified as semifusinite having cell lumen with clay mineral infillings. In Fig. 4.8, the inertinite macerals found in the Rajmahal shale and coal samples with partially preserved cell structure is semifusinite (Fig. 4.8a: ROI-1, b: ROI-2, respectively). When the marked inertinite macerals were moved under SEM using the correlative technique, it was observed that the semifusinite found in the coal sample had cell lumens with clay mineral infillings (Fig. 4.8d: EDS Spot 1 and f) while the semifusinite found in the shale sample did not (Fig. 4.8c, e). Similar observations were made by Sethi et al. (2023a, b). A

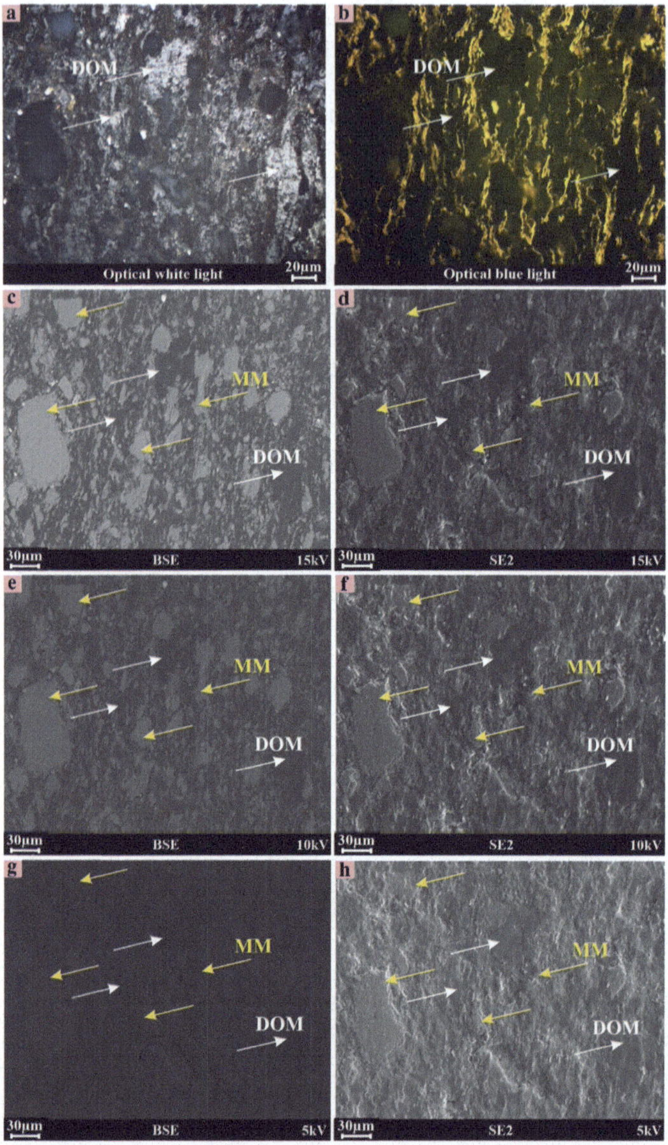

Fig. 4.4 Correlative image set of shale sample from Rajmahal Basin, India, under **a** brightfield reflected light, **b** fluorescence light, **c** 15 kV BSE at 750× magnification, **d** 15 kV SE2 at 750× magnification, **e** 10 kV BSE at 750× magnification, **f** 10 kV SE2 at 750× magnification, **g** 5 kV BSE at 750× magnification, and **h** 5 kV SE2 at 750× magnification. *Abbreviations* DOM—dispersed organic matter; MM—mineral matter

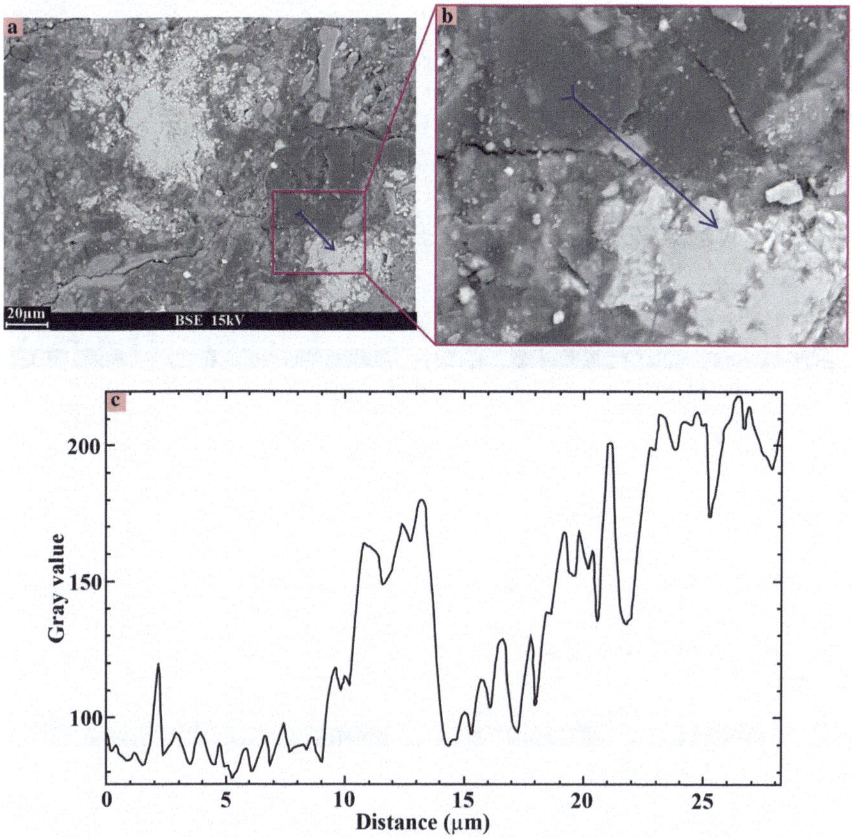

Fig. 4.5 BSE image gray-value profile plot of the shale sample from Rajmahal Basin

key SEM distinguishing feature in some inertinites is the presence of cellular structures made of vascular plant cell lumens (Liu et al. 2022; Sethi et al. 2023a). In the absence of such features, it may be challenging to determine the maceral type using SEM alone. For example, in Fig. 4.9, the inertinite maceral observed during optical microscopy (Fig. 4.9a: ROI-1) would be difficult to identify under SEM alone as it does not exhibit any distinctive feature (Fig. 4.9c, e). Similarly, the inertinite and vitrinite maceral adjacent to each other (Fig. 4.9b) were clearly differentiable during optical microscopy, while under SEM there is no discernible difference between the two that can be used to identify them without the use of the correlative technique (Fig. 4.9e, f). Although grayscale contrast can be seen between the vitrinite and inertinite macerals in Rajmahal coal under SEM in BSE scanning mode (Fig. 4.9d), it is not sufficiently noteworthy. Cardott and Curtis (2018) discovered that grayscale contrast alone was insufficient to differentiate between the inertinite macerals (semifusinite, fusinite, and macrinite) with the exception when the maceral possessed a cellular structure (fusinite vs. semifusinite). The distinction between inertinite and

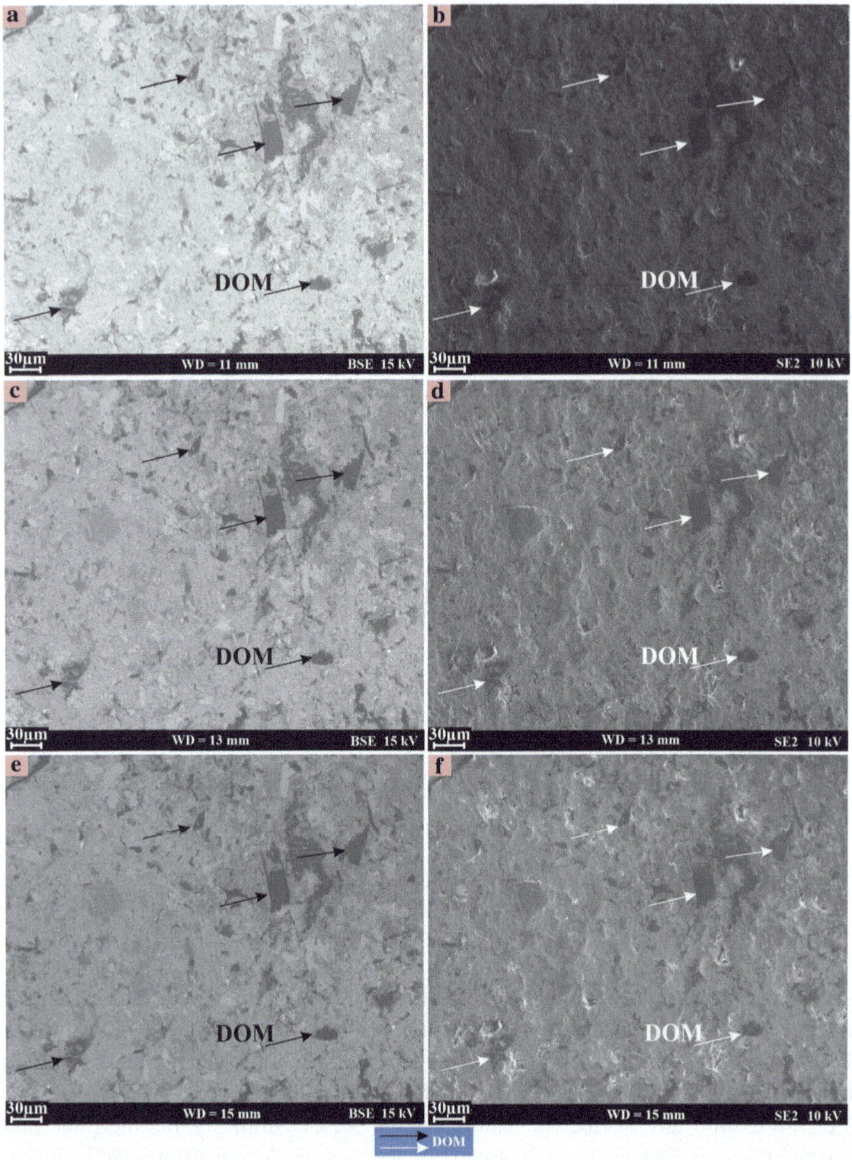

Fig. 4.6 SEM photomicrographs of the shale sample at different working distances: 11 mm (**a**, **b**), 13 mm (**b**, **c**), and 15 mm (**c**, **d**). *Abbreviations* DOM—dispersed organic matter. Reprinted from International Journal of Coal Geology, Vol. 274, Sethi et al. (2023b), Using optical-electron correlative microscopy for shales of contrasting thermal maturity, Page No. 104273, Copyright (2023), with permission from Elsevier

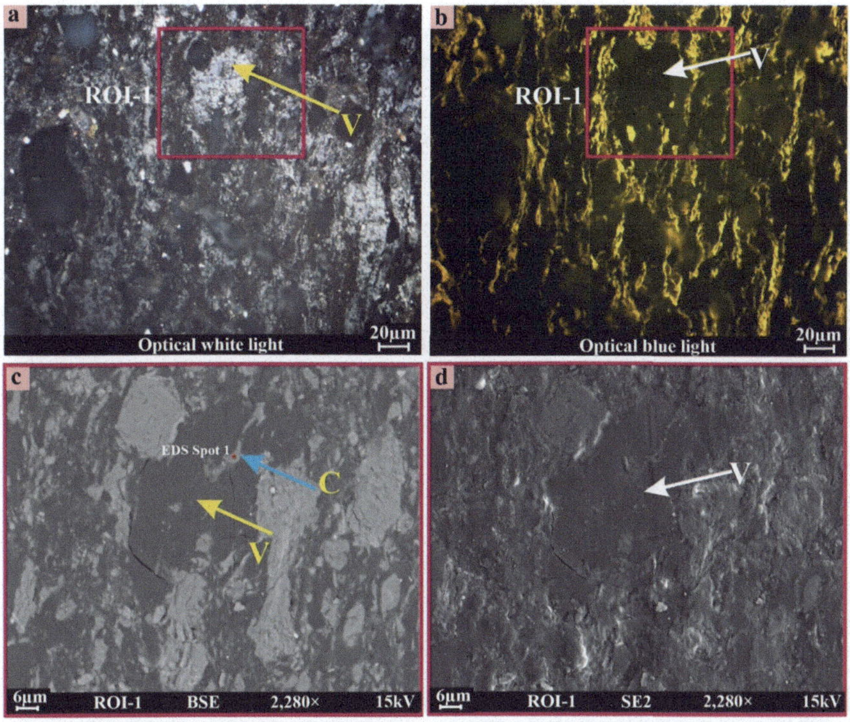

Fig. 4.7 Correlative image set of vitrinite found in Rajmahal shale sample. *Abbreviations* V—vitrinite; C—clay; ROI—region of interest

vitrinite in shale samples was also applied, and it was found that there was little to grayscale contrast between the two (Sethi et al. 2023a, b). Figures 4.7 and 4.8 show that the vitrinite (Fig. 4.7d) and inertinite (Figs. 4.7d and 4.8c) macerals, respectively, have negative relief in comparison to the surrounding mineral matrix. On the other hand, it was observed that inertinite maceral showing positive relief compared to vitrinite (Figs. 4.8f and 4.9f). Relief serves as a tool for recognizing potential woody terrigenous organic materials (Loucks and Reed 2014; Yang et al. 2016). However, precise identification necessitates correlative microscopy, especially when the sample undergoes ion milling, as it lacks relief to differentiate between various organic matter types (Valentine and Hackley 2019).

Figure 4.10 shows the correlative image set of liptinite macerals (Fig. 4.10a–i) observed in the shale samples from Rajmahal and Jharia basins of India. Sporinite found in the Rajmahal shale samples appear dark under the SEM in BSE scanning mode at 15 kV acceleration voltage (Fig. 4.10b, e). Sporinite is easy to identify under SEM due to its distinct morphology. Of the three maceral groups, liptinite macerals (Pickel et al. 2017) are relatively easy to identify under SEM due to their

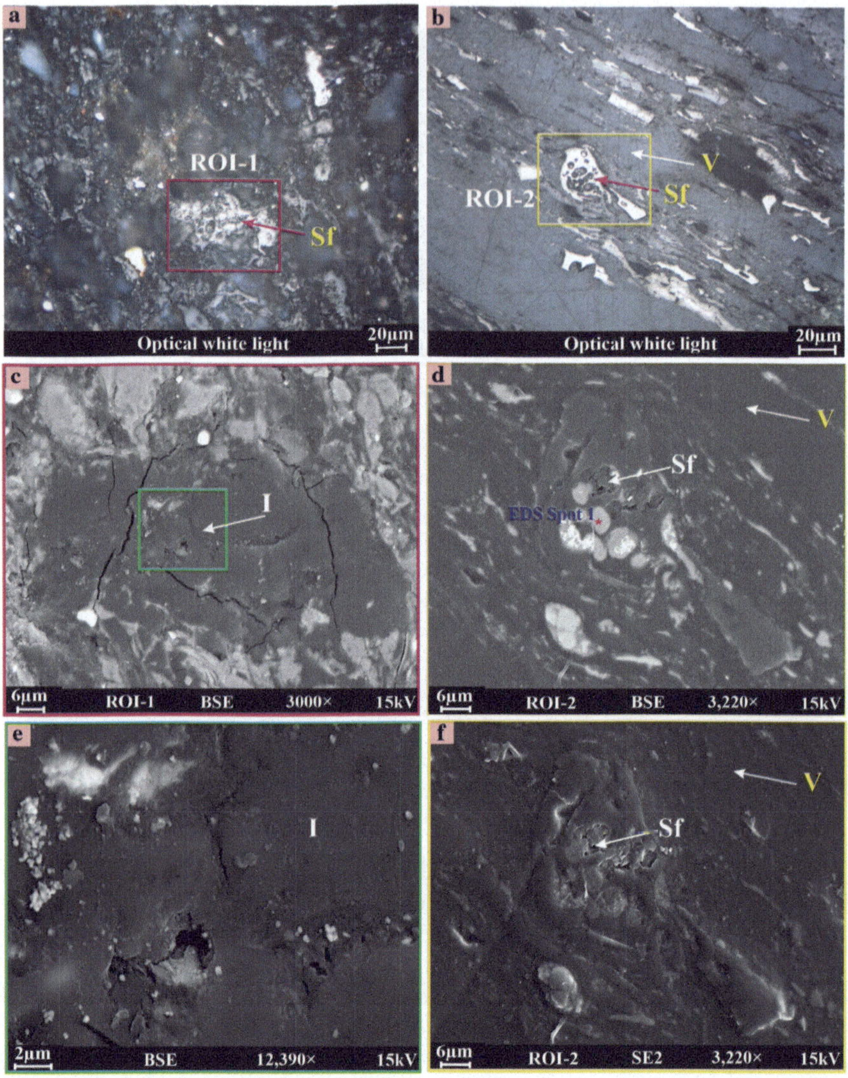

Fig. 4.8 Correlative image set of semifusinite and vitrinite found in Rajmahal shale (**a**, **c**, and **e**) and Jharia coal (**b**, **d**, and **f**) sample. *Abbreviations* Sf—semifusinite; V—vitrinite; I—inertinite; ROI—region of interest

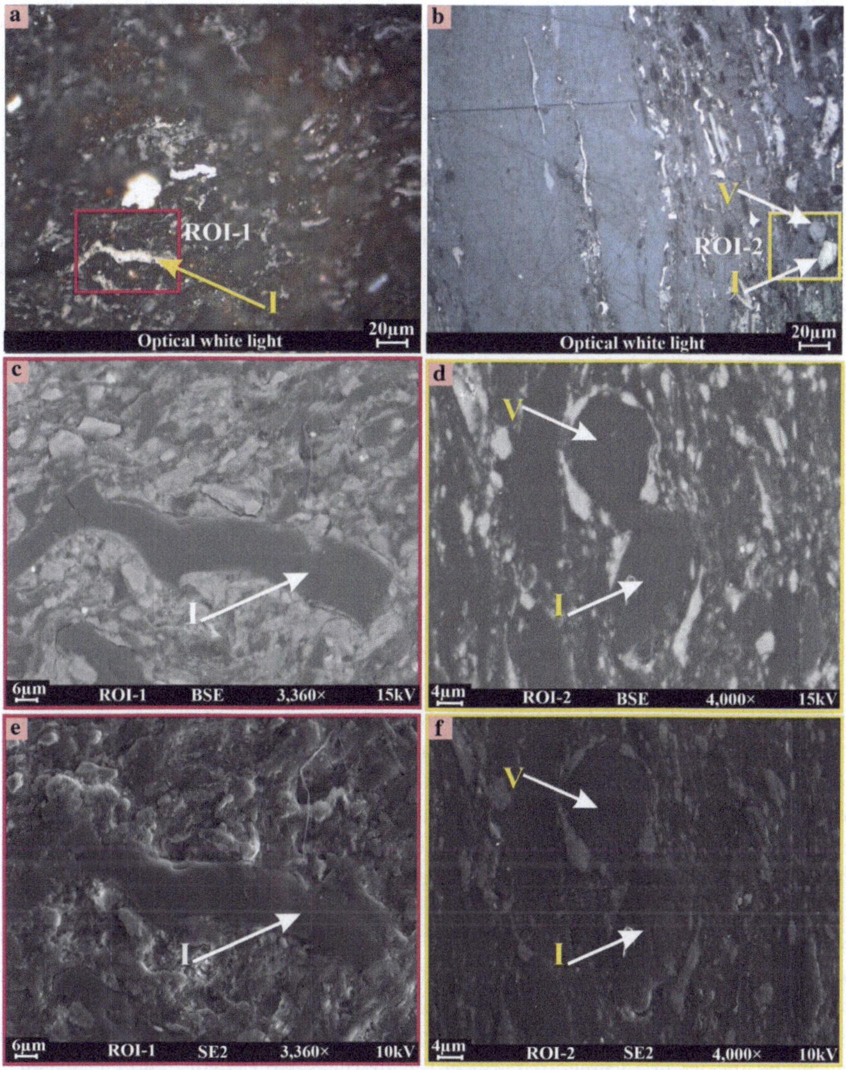

Fig. 4.9 Correlative image set of inertinite and vitrinite found in Rajmahal shale (**a**, **c**, and **e**) and Jharia coal (**b**, **d**, and **f**) sample. *Abbreviations* V—vitrinite; I—inertinite; ROI—region of interest

distinct dark gray to black color and distinctive structure (Sethi et al. 2023b). In Fig. 4.11, resinite exhibiting greater relief in contrast to the surrounding mineral matrix (Fig. 4.11f). In a correlative study conducted by Sethi et al. (2023b) it was observed that the liptinite macerals found (Fig. 4.11a, d) in thermally immature shales had positive relief (Fig. 4.11b, c, e, f) whereas negative relief was observed for liptinite macerals found in thermally late-mature-stage shales (Fig. 4.11g–i). The low hardgrove grindability index (HGI) of coal influenced by liptinite as shown by several studies (Hower et al. 1987; Trimble and Hower 2003; Rejdak et al. 2021) has been defined as an indication of liptinites as hard coal constituents that increase toughness and provide resistance to grinding forces. Thus, liptinites may resist the mechanical polishing during the process of shale pellet preparation for correlative analysis. The negative relief observed for liptinites found in higher maturity levels in shales can stem from alterations associated with hydrocarbon generation, impacting not only their chemical and optical attributes but also their physical characteristics. Sporinite found in Rajmahal shales (Fig. 4.10a: ROI-1, b: ROI-2) were observed to have SEM visible pores (Fig. 4.10c, f), whereas no pores were visible (Fig. 4.10h, i) in the liptinite macerals found in the Jharia shales (Fig. 4.10g: ROI-1). It should also be noted that in case of vitrinite maceral, no pores were found in the shale and coal samples (Figs. 4.7d, 4.8f and 4.9f), whereas primary porosity was observed in inertinite maceral with or without mineral infillings (Figs. 4.8e and 4.9f, respectively). Similar observations were made in the correlative study conducted by Sethi et al. (2023b) that compared the development of organic matter-hosted porosity in the thermally contrasting shales of Rajmahal and Jharia Basins. They observed 200–300 nm sized pores in sporinite found in thermally immature shales of Rajmahal Basin (Fig. 4.11: ROI-1, B and C). It was attribute to the inherent characteristics of type-II kerogens, which are more inclined to generate hydrocarbons at low thermal maturity levels compared to type-III kerogens (Jarvie et al. 2007). Although, the Jharia Basin shales studied by Sethi et al. (2023b) were at late mature stage of thermal maturity (at the boundary of the oil window and condensate-wet gas window), no pores were visibile in the liptinite (Fig. 4.11g: ROI-3, h, i) macerals. They explained that the possible reasons could be: (i) predominantly microporous (< 2 nm) character of high maturity coals making it difficult to see porosity under SEM (Clarkson and Bustin 1996); (ii) contact metamorphism (Varga and Horvath 1986; Mastalerz et al. 2009) caused by the intruded igneous body in the Jharia Basin (Mishra and Cook 1992; Singh et al. 2008, 2013); (iii) rearrangement of the structured organic matter due to aromatization during the expulsion or diffusion of oil produced from its structure (Inan et al. 2018). In several past studies that applied correlative technique (Valentine and Hackley 2019; Liu et al. 2022) it was observed that Porosity hosted within the organic matter is primarily identified in the solid bitumen within the shale sample, rather than in the kerogen. Solid bitumen may also undergo aromatic rearrangement (Curiale 1986), but it is more probable that the bitumen migrates into the pore space between rigid minerals, preventing additional compaction. The fact that DOM is dispersed and/or layered within the source rock supports the idea that it is a load-bearing material (Inan et al. 2018). DOM must therefore go through a subsequent stage of compaction during the oil/gas release process, which could result in it losing

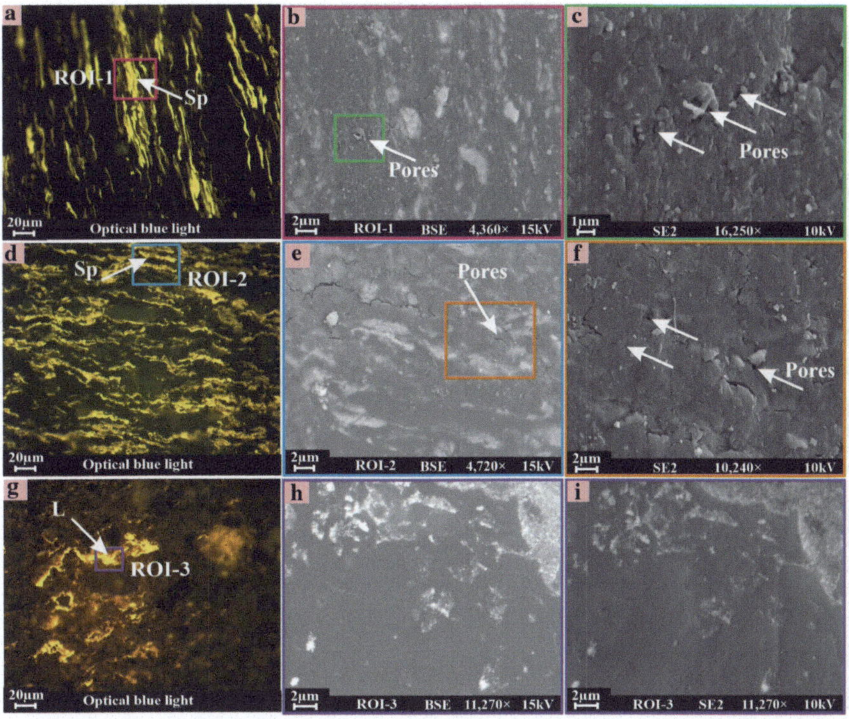

Fig. 4.10 Correlative image set of liptinite macerals found in Rajmahal coal and shale (**a–f**) and Jharia shale samples (**g–i**). *Abbreviations* Sp—sporinite; L—liptinite; ROI—region of interest

its previously developed organic porosity. Consequently, during the oil/gas release process, DOM must go through a subsequent stage of compaction, which could result in the loss of its previously developed organic porosity.

In reflected light microscopy, identifying mineral matter can be challenging. For example, results of EDS analysis under SEM revealed presence of quartz in the Rajmahal shale sample (Fig. 4.12b: EDS Spot 1; c) which was previously not identifiable during optical microscopy (Fig. 4.12a). In Fig. 4.12, a bright white mass was observed in the Rajmahal shale sample during optical microscopy (Fig. 4.12d: ROI-2). Relocation of the marked position under SEM using the correlative technique and subsequent EDS analysis revealed it to be ilmenite (Fig. 4.12e: EDS Spot 2; f). Similarly, masses with very high reflectance were observed during the optical microscopy of Jharia shale samples (Fig. 4.12g: ROI-3). Relocation of the marked position under SEM followed byEDS analysis revealed it to be rutile (Fig. 4.12h, i). Interestingly, at higher magnification of 8250× under BSE scanning mode quartz was observed to be associated with rutile (Fig. 4.12i). Sethi et al. (2023b) in their study of Rajmahal shales using correlative microscopy found association of pyrite within the

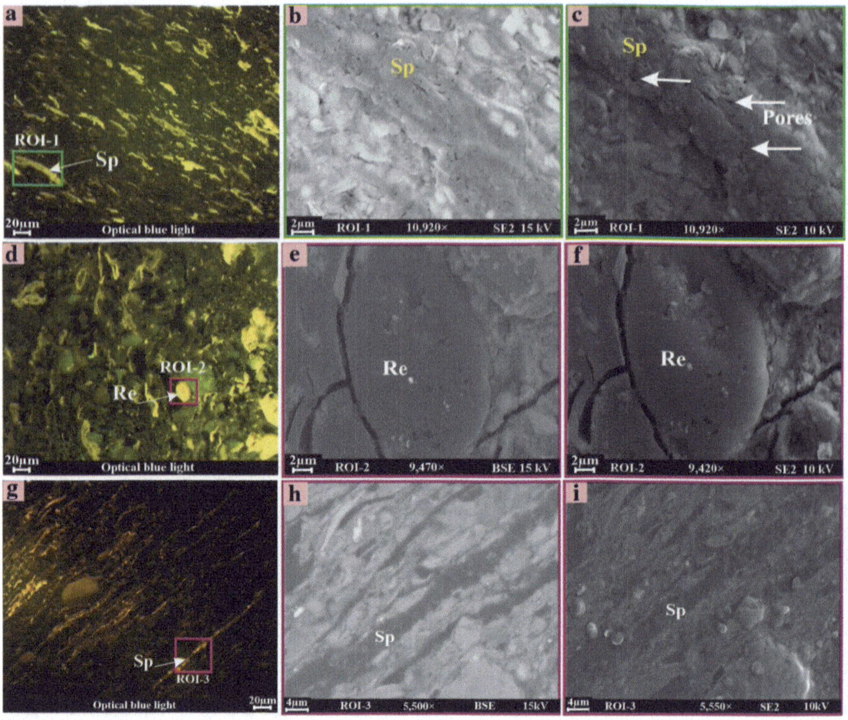

Fig. 4.11 Correlative image set of liptinite macerals found in Rajmahal (**a**–**f**) and Jharia shale samples (**g**–**i**). *Abbreviations* Sp—sporinite; Re—resinite; ROI—region of interest. Reprinted from International Journal of Coal Geology, Vol. 274, Sethi et al. (2023b), Using optical-electron correlative microscopy for shales of contrasting thermal maturity, Page No. 104273, Copyright (2023), with permission from Elsevier

vitrinite maceral under SEM (Fig. 4.13b, c) which was not visible during the optical microscopy of the same position (Fig. 4.13a: ROI-1). Additionally, they noticed interstitial spaces between pyrite grains found in Jharia shale samples (Fig. 4.13d: ROI-2) showing fluorescence under blue light illumination (Fig. 4.13d). Relocation of the marked position under SEM (Fig. 4.13f), and subsequent EDS spot analysis of the interstitial spaces revealed presence of C, Al, Si. In Fig. 4.13, bright white masses were observed in the coal samples of Rajmahal and Jharia basins (Fig. 4.13g: ROI-3, j: ROI-4, respectively) under the optical microscope. The marked positions were moved under the electron microscope employing the correlative technique. White masses found in the Rajmahal coal sample are observed to be bright under SEM and the following EDS analysis revealed it to be pyrite (Fig. 4.13h, i). Contrarily, the white mass in the Jharia coal sample is dark in appearance and was identified as organic matter by EDS analysis (Fig. 4.13k, l). The higher reflectance of the organic matter (Fig. 4.13j: ROI-4) observed during optical microscopy could be attributed to graphitization due to contact metamorphism (Buseck and Beyssac 2014). Without the

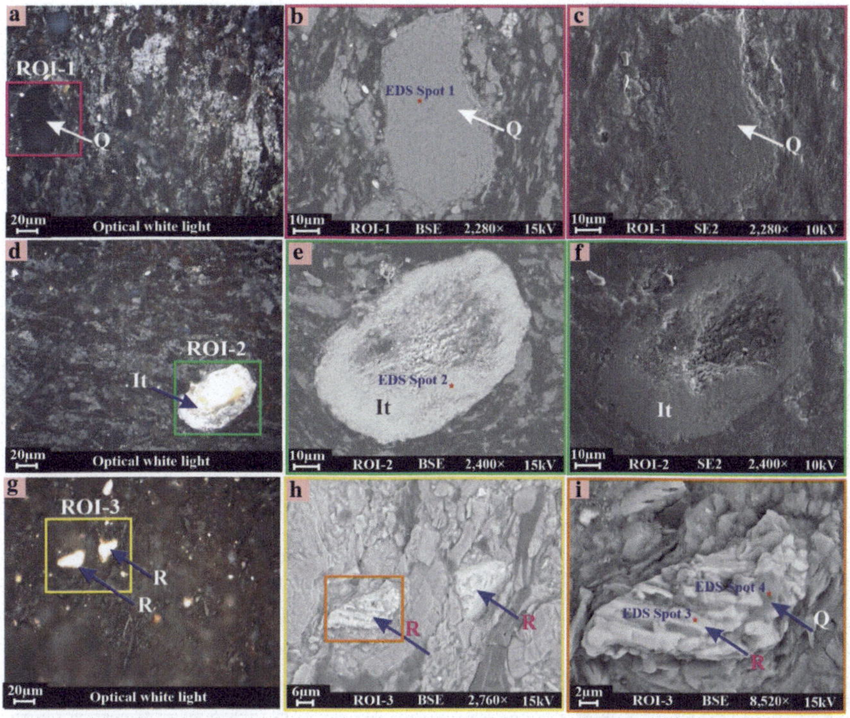

Fig. 4.12 Correlative image set of minerals found in Rajmahal shale samples. *Abbreviations* Q—quartz; It—ilmenite; R—rutile; ROI—region of interest

correlative technique it would not have been possible to make these important observations under optical microscope alone. For instance, the organic matter's association with pyrite is of great significance because it can increase the alkane production while lowering the activation energy required for the production of hydrocarbons form kerogen, increasing hydrocarbon generation from kerogen by 20–50% as a result of catalytic activity (Ma et al. 2016).

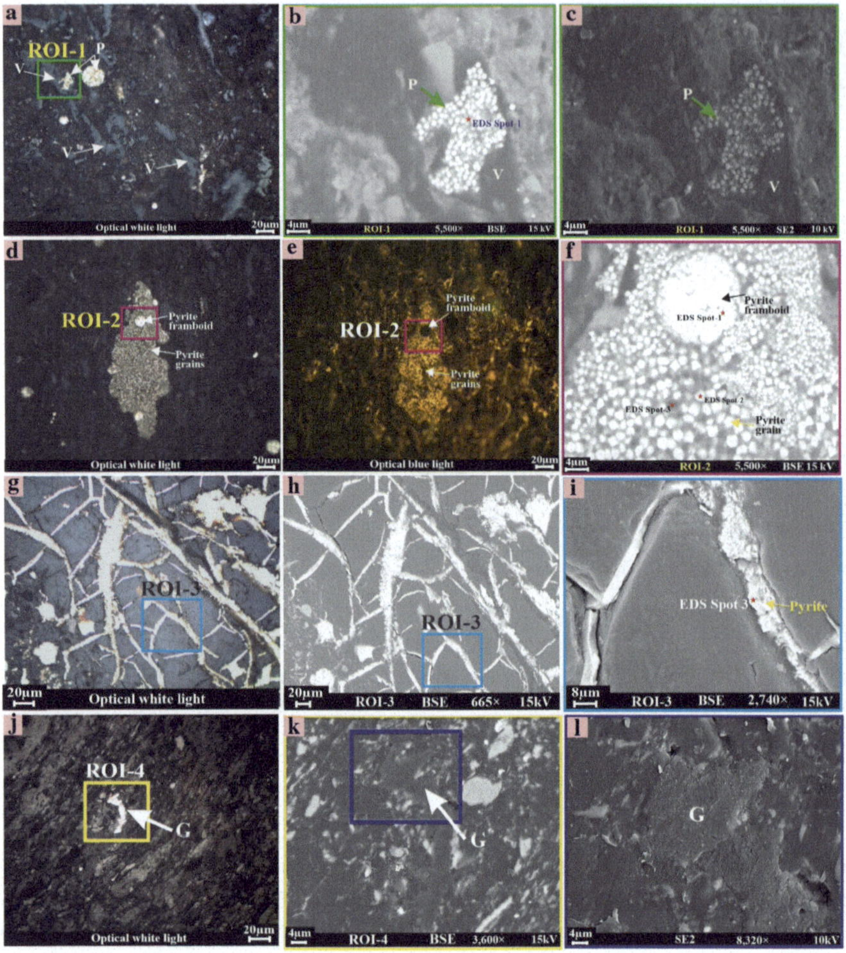

Fig. 4.13 Correlative image set of Rajmahal and Jharia shale (**a**–**f**) and coal (**g**–**l**) samples for assessment of their mineralogical composition using the correlative method. *Abbreviations* P—pyrite; V—vitrinite; G—graphite. Images **a**–**f** reprinted from International Journal of Coal Geology, Vol. 274, Sethi et al. (2023b), Using optical-electron correlative microscopy for shales of contrasting thermal maturity, Page No. 104273, Copyright (2023), with permission from Elsevier

References

Abarghani A, Gentzis T, Shokouhimehr M, Liu B, Ostadhassan M (2020) Chemical heterogeneity of organic matter at nanoscale by AFM-based IR spectroscopy. Fuel 261:116454. https://doi.org/10.1016/j.fuel.2019.116454

Allen JA, Glenn M, Hapugoda P, Stanger R, O'Brien G, Donne SW (2018) An investigation of mineral distribution in coking and thermal coal chars as fuels for the direct carbon fuel cell. Fuel 217:11–20. https://doi.org/10.1016/J.FUEL.2017.12.084

Arif M, Mahmoud M, Zhang Y, Iglauer S (2021) X-ray tomography imaging of shale microstructures: a review in the context of multiscale correlative imaging. Int J Coal Geol 233:103641. https://doi.org/10.1016/J.COAL.2020.103641

Asadi P, Beckingham LE (2021) Integrating machine/deep learning methods and filtering techniques for reliable mineral phase segmentation of 3D X-ray computed tomography images. Energies 14:4595. https://doi.org/10.3390/EN14154595

Bernard S, Horsfield B, Schulz H-M, Wirth R, Schreiber A, Sherwood N (2012a) Geochemical evolution of organic-rich shales with increasing maturity: a STXM and TEM study of the Posidonia Shale (lower Toarcian, northern Germany). Mar Pet Geol 31:70–89. https://doi.org/10.1016/j.marpetgeo.2011.05.010

Bernard S, Wirth R, Schreiber A, Schulz H-M, Horsfield B (2012b) Formation of nanoporous pyrobitumen residues during maturation of the Barnett Shale (Fort Worth Basin). Int J Coal Geol 103:3–11. https://doi.org/10.1016/j.coal.2012.04.010

Bernard S, Brown L, Wirth R, Schreiber A, Schulz H-M, Horsfield B, Aplin AC, Mathia EJ (2013) FIB-SEM and TEM investigations of an organic-rich shale maturation series from the lower Toarcian Posidonia Shale, Germany <subtitle>Nanoscale Pore System and Fluid-rock Interactions</subtitle>. In: Electron microscopy of shale hydrocarbon reservoirs. https://doi.org/10.1306/13391705m1023583

Boutchko R, Rayz VL, Vandehey NT, O'Neil JP, Budinger TF, Nico PS, Druhan JL, Saloner DA, Gullberg GT, Moses WW (2012) Imaging and modeling of flow in porous media using clinical nuclear emission tomography systems and computational fluid dynamics. J Appl Geophy 76:74–81. https://doi.org/10.1016/j.jappgeo.2011.10.003

Brattekås B, Steinsbø M, Graue A, Fernø MA, Espedal H, Seright RS (2017) New insight into wormhole formation in polymer gel during water chase floods with positron emission tomography. SPE J 22:32–40. https://doi.org/10.2118/180051-pa

Buchardt B, Lewan MD (1990) Reflectance of Vitrinite-like macerals as a thermal maturity index for Cambrian-Ordovician Alum Shale, Southern Scandinavia. Am Assoc Pet Geol Bull 74. https://doi.org/10.1306/0c9b230d-1710-11d7-8645000102c1865d

Bultreys T, De Boever W, Cnudde V (2016) Imaging and image-based fluid transport modeling at the pore scale in geological materials: a practical introduction to the current state-of-the-art. Earth Sci Rev 155:93–128. https://doi.org/10.1016/j.earscirev.2016.02.001

Buseck PR, Beyssac O (2014) From organic matter to graphite: graphitization. Elements 10(6):421–426

Cai Y, Liu D, Mathews JP, Pan Z, Elsworth D, Yao Y, Li J, Guo X (2014) Permeability evolution in fractured coal—combining triaxial confinement with X-ray computed tomography, acoustic emission and ultrasonic techniques. Int J Coal Geol 122:91–104. https://doi.org/10.1016/J.COAL.2013.12.012

Cała M, Cyran K, Stopkowicz A, Kolano M, Szczygielski M (2016) Preliminary application of X-ray computed tomograph on characterisation of polish gas shale mechanical properties. Rock Mech Rock Eng 49:4935–4943. https://doi.org/10.1007/S00603-016-1045-6/FIGURES/4

Camp WK (2016) Strategies for identifying organic matter types in SEM. In: SEPM-AAPG Hedberg research mudstone diagenesis conference 16–19

Chandra D, Vishal V (2020) A comparison of nano-scale pore attributes of Barakar formation gas shales from Raniganj and Wardha Basin, India using low pressure sorption and FEG-SEM analysis. J Nat Gas Sci Eng 81:103453. https://doi.org/10.1016/J.JNGSE.2020.103453

Chandra D, Vishal V (2021) A critical review on pore to continuum scale imaging techniques for enhanced shale gas recovery. Earth Sci Rev 217:103638. https://doi.org/10.1016/J.EARSCIREV.2021.103638

Chandra D, Vishal V (2022) A comparative analysis of pore attributes of sub-Bituminous Gondwana Coal from the Damodar and Wardha Valleys: implication for enhanced coalbed methane recovery. Energy Fuels. https://doi.org/10.1021/ACS.ENERGYFUELS.2C00854/ASSET/IMAGES/LARGE/EF2C00854_0013.JPEG

Chandra D, Vishal V, Bahadur J, Sen D (2020a) A novel approach to identify accessible and inaccessible pores in gas shales using combined low-pressure sorption and SAXS/SANS analysis. Int J Coal Geol 228:103556. https://doi.org/10.1016/J.COAL.2020.103556

Chandra D, Bakshi T, Vishal V (2021) Thermal effect on pore characteristics of shale under inert and oxic environments: insights on pore evolution. Microporous Mesoporous Mater 316:110969. https://doi.org/10.1016/J.MICROMESO.2021.110969

Chandra D, Bakshi T, Bahadur J, Hazra B, Vishal V, Kumar S, Sen D, Singh TN (2023) Pore morphology in thermally-treated shales and its implication on CO_2 storage applications: a gas sorption, SEM, and small-angle scattering study. Fuel 331:125877. https://doi.org/10.1016/J.FUEL.2022.125877

Chandra D, Vishal V, Debbarma A, Banerjee S, Pradhan SP, Mishra MK (2020b) Role of composition and depth on pore attributes of Barakar Formation gas shales of Ib Valley, India using a combination of low pressure sorption and image analysis. Energy Fuels. https://doi.org/10.1021/acs.energyfuels.0c00746

Cheng P, Zhang CP, Ma ZY, Zhou JP, Zhang DC, Liu XF, Chen H, Ranjith PG (2022) Experimental study of micromechanical properties alterations of shale matrix treated by $ScCO_2$-water saturation using nanoindentation tests. Energy 242:122965

Cheng Y, Zhang X, Lu Z, Pan Z, Zeng M, Du X, Xiao S (2021) The effect of subcritical and supercritical CO_2 on the pore structure of bituminous coals. J Nat Gas Sci Eng 94:104132. https://doi.org/10.1016/J.JNGSE.2021.104132

Clarkson CR, Bustin RM (1996) Variation in micropore capacity and size distribution with composition in bituminous coal of the Western Canadian Sedimentary Basin: implications for coalbed methane potential. Fuel 75(13):1483–1498

Curiale JA (1986) Origin of solid bitumens, with emphasis on biological marker results. Org Geochem 10(1–3):559–580

Dai X, Wang M, Wei C, Zhang J, Wang X, Zou M (2020) Factors affecting shale microscopic pore structure variation during interaction with supercritical CO_2. J CO_2 Utilization 38:194–211. https://doi.org/10.1016/J.JCOU.2020.01.021

Davudov D, Moghanloo RG, Zhang Y (2020) Interplay between pore connectivity and permeability in shale sample. Int J Coal Geol 220:103427. https://doi.org/10.1016/J.COAL.2020.103427

Duan Y, Feng XT, Li X, Yang B (2022) Mesoscopic damage mechanism and a constitutive model of shale using in-situ X-ray CT device. Eng Fract Mech 269:108576. https://doi.org/10.1016/J.ENGFRACMECH.2022.108576

Duber S (2011) Nanostruktura i mikrotekstura antracytów

Fatah A, Ben Mahmud H, Bennour Z, Gholami R, Hossain M (2022) The impact of supercritical CO_2 on the pore structure and storage capacity of shales. J Nat Gas Sci Eng 98:104394. https://doi.org/10.1016/J.JNGSE.2021.104394

Fernø MA, Gauteplass J, Hauge LP, Abell GE, Adamsen TCH, Graue A (2015) Combined positron emission tomography and computed tomography to visualize and quantify fluid flow in sedimentary rocks. Water Resour Res 51:7811–7819. https://doi.org/10.1002/2015wr017130

Ganguly E, Misra S, Wu Y (2020) Generalizable data-driven techniques for microstructural analysis of shales. In: Proceedings—SPE annual technical conference and exhibition 2020-October. https://doi.org/10.2118/201554-MS

Garum M, Glover PWJ, Lorinczi P, Drummond-Brydson R, Hassanpour A (2020) Micro- and nano-scale pore structure in gas shale using Xμ-CT and FIB-SEM techniques. Energy Fuels 34:12340–12353. https://doi.org/10.1021/ACS.ENERGYFUELS.0C02025/ASSET/IMAGES/LARGE/EF0C02025_0015.JPEG

Gu X, Cole DR, Rother G, Mildner DFR, Brantley SL (2015) Pores in marcellus shale: a neutron scattering and FIB-SEM study. Energy Fuels 29:1295–1308. https://doi.org/10.1021/ACS.ENERGYFUELS.5B00033/ASSET/IMAGES/LARGE/EF-2015-00033G_0005.JPEG

Hackley PC, Valentine BJ, Voortman LM, Van Oosten Slingeland DS, Hatcherian J (2017) Utilization of integrated correlative light and electron microscopy (iCLEM) for imaging sedimentary organic matter. J Microsc 267(3):371–383

Hazra B, Chandra D, Lahiri S, Vishal V, Sethi C, Pandey JK (2023) Pore evolution during combustion of distinct thermally mature shales: insights into potential in situ conversion. Energy Fuels 37(18):13898–13911

Hower JC, Graese AM, Klapheke JG (1987) Influence of microlithotype composition on Hardgrove grindability for selected Eastern Kentucky coals. Int J Coal Geol 7(3):227–244

Inan S, Al Badairy H, Inan T, Al Zahrani A (2018) Formation and occurrence of organic matter-hosted porosity in shales. Int J Coal Geol 199:39–51

Javadpour F, Moravvej Farshi M, Amrein M (2012) Atomic-force microscopy: a new tool for gas-shale characterization. J Can Pet Technol 51:236–243. https://doi.org/10.2118/161015-pa

Joss L, Pini R (2018) 3D mapping of gas physisorption for the spatial characterisation of nanoporous materials. ChemPhysChem. https://doi.org/10.1002/cphc.201801148

Kazak A, Simonov K, Kulikov V (2021) Machine-learning-assisted segmentation of focused ion beam-scanning electron microscopy images with artifacts for improved void-space characterization of tight reservoir rocks. SPE J 26:1739–1758. https://doi.org/10.2118/205347-PA

Kelly S, El-Sobky H, Torres-Verdín C, Balhoff MT (2016) Assessing the utility of FIB-SEM images for shale digital rock physics. Adv Water Resour 95:302–316. https://doi.org/10.1016/J.ADVWATRES.2015.06.010

King HE, Eberle APR, Walters CC, Kliewer CE, Ertas D, Huynh C (2015) Pore architecture and connectivity in gas shale. Energy Fuels 29:1375–1390. https://doi.org/10.1021/EF502402E/ASSET/IMAGES/LARGE/EF-2014-02402E_0009.JPEG

Knaup AS, Jernigen JD, Curtis ME, Sholeen JW, Borer JJ, Sondergeld CH, Rai CS (2019) Unconventional reservoir microstructural analysis using sem and machine learning. In: SPE/AAPG/SEG unconventional resources technology conference 2020, URTeC 2020. https://doi.org/10.15530/URTEC-2019-638

Kulenkampff J, Gründig M, Richter M, Enzmann F (2008) Evaluation of positron-emission-tomography for visualisation of migration processes in geomaterials. Phys Chem Earth, Parts a/b/c 33:937–942. https://doi.org/10.1016/j.pce.2008.05.005

Kumar S, Das S, Bastia R, Ojha K (2018) Mineralogical and morphological characterization of Older Cambay Shale from North Cambay Basin, India: implication for shale oil/gas development. Mar Pet Geol 97:339–354. https://doi.org/10.1016/j.marpetgeo.2018.07.020

Kwiecińska B, Petersen HI (2004) Graphite, semi-graphite, natural coke, and natural char classification ICCP system. Int J Coal Geol 57:99–116. https://doi.org/10.1016/j.coal.2003.09.003

Kwiecińska B, Pusz S, Valentine BJ (2019) Application of electron microscopy TEM and SEM for analysis of coals, organic-rich shales and carbonaceous matter. Int J Coal Geol 211:103203. https://doi.org/10.1016/j.coal.2019.05.010

Li Z, Liu D, Cai Y, Ranjith PG, Yao Y (2017) Multi-scale quantitative characterization of 3-D pore-fracture networks in bituminous and anthracite coals using FIB-SEM tomography and X-ray μ-CT. Fuel 209:43–53. https://doi.org/10.1016/J.FUEL.2017.07.088

Li C, Ostadhassan M, Guo S, Gentzis T, Kong L (2018a) Application of PeakForce tapping mode of atomic force microscope to characterize nanomechanical properties of organic matter of the Bakken Shale. Fuel 233:894–910. https://doi.org/10.1016/j.fuel.2018.06.021

Li Z, Fu Y, Yang C, Yu W, Liu L, Qu J, Zhao W (2018b) Mineral liberation analysis on coal components separated using typical comminution methods. Miner Eng 126:74–81. https://doi.org/10.1016/J.MINENG.2018.06.028

Li S, Liu L, Chai P, Li X, He J, Zhang Z, Wei L (2019) Imaging hydraulic fractures of shale cores using combined positron emission tomography and computed tomography (PET-CT) imaging technique. J Pet Sci Eng. https://doi.org/10.1016/j.petrol.2019.106283

Li Q, Liu D, Cai Y, Zhao B, Qiu Y, Zhou Y (2020) Scale-span pore structure heterogeneity of high volatile bituminous coal and anthracite by FIB-SEM and X-ray μ-CT. J Nat Gas Sci Eng 81:103443. https://doi.org/10.1016/J.JNGSE.2020.103443

Li B, Nie X, Cai J, Zhou X, Wang C, Han D (2022a) U-Net model for multi-component digital rock modeling of shales based on CT and QEMSCAN images. J Pet Sci Eng 216:110734. https://doi.org/10.1016/J.PETROL.2022.110734

Li Y, Chen J, Elsworth D, Pan Z, Ma X (2022b) Nanoscale mechanical property variations concerning mineral composition and contact of marine shale. Geosci Front 13:101405. https://doi.org/10.1016/J.GSF.2022.101405

Li B, Jian W, Zhang J, Wang B, Zhu D (2023) Analysis of the change feature and action mechanism of coal pore structure under the action of supercritical CO_2. Int J Energy Res 2023. https://doi.org/10.1155/2023/8679917

Lin S, Hou L, Luo X (2022) Shale mineralogy analysis method: quantitative correction of minerals using QEMSCAN based on MAPS technology. Appl Sci (switzerland) 12:5013. https://doi.org/10.3390/APP12105013/S1

Lippmann-Pipke J, Gerasch R, Schikora J, Kulenkampff J (2017) Benchmarking PET for geoscientific applications: 3D quantitative diffusion coefficient determination in clay rock. Comput Geosci 101:21–27. https://doi.org/10.1016/j.cageo.2017.01.002

Liu S, Sang S, Wang G, Ma J, Wang X, Wang W, Du Y, Wang T (2017) FIB-SEM and X-ray CT characterization of interconnected pores in high-rank coal formed from regional metamorphism. J Pet Sci Eng 148:21–31. https://doi.org/10.1016/J.PETROL.2016.10.006

Liu S, Wang Z, Zhang L (2018) Experimental study on the cracking process of layered shale using X-ray microCT. Energy Explor Exploit 36:297–313. https://doi.org/10.1177/0144598717736855/ASSET/IMAGES/LARGE/10.1177_0144598717736855-FIG12.JPEG

Liu X, Nie B, Wang W, Wang Z, Zhang L (2019a) The use of AFM in quantitative analysis of pore characteristics in coal and coal-bearing shale. Mar Pet Geol 105:331–337. https://doi.org/10.1016/j.marpetgeo.2019.04.021

Liu X, Song D, He X, Wang Z, Zeng M, Wang L (2019b) Quantitative analysis of coal nanopore characteristics using atomic force microscopy. Powder Technol 346:332–340. https://doi.org/10.1016/j.powtec.2019.02.027

Liu B, Mastalerz M, Schieber J (2022) SEM petrography of dispersed organic matter in black shales: a review. Earth Sci Rev 224:103874

Liu Q, Sun M, Sun X, Liu B, Ostadhassan M, Huang W, Chen X, Pan Z (2023) Pore network characterization of shale reservoirs through state-of-the-art X-ray computed tomography: a review. Gas Sci Eng 113:204967. https://doi.org/10.1016/J.JGSCE.2023.204967

Loucks RG, Reed RM (2014) Scanning-electron-microscope petrographic evidence for distinguishing organic-matter pores associated with depositional organic matter versus migrated organic matter in mudrocks. Gulf Coast Assoc Geol Soc J 3:51–60

Loucks RG, Ruppel SC (2007) Mississippian Barnett Shale: Lithofacies and depositional setting of a deep-water shale-gas succession in the Fort Worth Basin, Texas. Am Assoc Pet Geol Bull 91:579–601. https://doi.org/10.1306/11020606059

Loucks RG, Reed RM, Ruppel SC, Jarvie DM (2009) Morphology, genesis, and distribution of nanometer-scale pores in siliceous mudstones of the Mississippian Barnett Shale. J Sediment Res 79(12):848–861

Ma X, Zheng J, Zheng G, Xu W, Qian Y, Xia Y, Wang Z, Wang X, Ye X (2016) Influence of pyrite on hydrocarbon generation during pyrolysis of type-III kerogen. Fuel 167:329–336

Mastalerz M, Drobniak A, Schimmelmann A (2009) Changes in optical properties, chemistry, and micropore and mesopore characteristics of bituminous coal at the contact with dikes in the Illinois Basin. Int J Coal Geol 77(3–4):310–319

Milliken KL, Ko LT, Pommer M, Marsaglia KM (2014) SEM petrography of eastern Mediterranean sapropels: analogue data for assessing organic matter in oil and gas shales. J Sediment Res 84(11):961–974

Milner M, McLin R, Petriello J (2010) Imaging texture and porosity in mudstones and shales: comparison of secondary and ion-milled backscatter SEM methods, SPE-138975. In: CSUG/SPE Canadian unconventional resources and international petroleum conference, Calgary. Alberta, Canada

Mishra HK, Cook AC (1992) Petrology and thermal maturity of coals in the Jharia Basin: implications for oil and gas origins. Int J Coal Geol 20(3–4):277–313

Pini R, Joss L (2019) See the unseen: applications of imaging techniques to study adsorption in microporous materials. Curr Opin Chem Eng 24:37–44. https://doi.org/10.1016/j.coche.2019.01.002

Pusz S, Borrego AG, Alvarez D, Camean I, du Cann V, Duber S, Kalkreuth W, Komorek J, Kus J, Kwiecińska BK, Libera M, Marques M, Misz-Kennan M, Morga R, Rodrigues S, Smędowski Ł, Suarez-Ruiz I, Strzezik J (2014) Application of reflectance parameters in the estimation of the structural order of coals and carbonaceous materials. Precision and bias of measurements derived from the ICCP structural working group. Int J Coal Geol 131:147–161. https://doi.org/10.1016/j.coal.2014.04.004

Rejdak M, Micorek T, Hower JC (2021) Influence of selected factors of polish coking coals on the Hardgrove Grindability Index (HGI). Int J Coal Prep Utilization 41(11):789–802

Saif T, Lin Q, Butcher AR, Bijeljic B, Blunt MJ (2017) Multi-scale multi-dimensional microstructure imaging of oil shale pyrolysis using X-ray micro-tomography, automated ultra-high resolution SEM, MAPS mineralogy and FIB-SEM. Appl Energy 202:628–647. https://doi.org/10.1016/J.APENERGY.2017.05.039

Sethi C, Hazra B, Wood DA, Singh AK (2023a) Experimental protocols to determine reliable organic geochemistry and geomechanical screening criteria for shales. J Earth Syst Sci 132(2):45

Sethi C, Mastalerz M, Hower JC, Hazra B, Singh AK, Vishal V (2023b) Using optical-electron correlative microscopy for shales of contrasting thermal maturity. Int J Coal Geol, p 104273

Singh AK, Sharma M, Singh MP (2008) Genesis of natural cokes: some Indian examples. Int J Coal Geol 75(1):40–48

Singh AK, Sharma M, Singh MP (2013) SEM and reflected light petrography: a case study on natural cokes from seam XIV, Jharia coalfield, India. Fuel 112:502–512

Song SB, Liu JF, Yang DS, Ni HY, Huang BX, Zhang K, Mao XB (2019) Pore structure characterization and permeability prediction of coal samples based on SEM images. J Nat Gas Sci Eng 67:160–171. https://doi.org/10.1016/J.JNGSE.2019.05.003

Sun M, Zhang L, Hu Q, Pan Z, Yu B, Sun L, Bai L, Fu H, Zhang Y, Zhang C, Cheng G (2020) Multiscale connectivity characterization of marine shales in southern China by fluid intrusion, small-angle neutron scattering (SANS), and FIB-SEM. Mar Pet Geol 112:104101. https://doi.org/10.1016/J.MARPETGEO.2019.104101

Tang X, Jiang Z, Jiang S, Li Z (2016) Heterogeneous nanoporosity of the Silurian Longmaxi formation shale gas reservoir in the Sichuan Basin using the QEMSCAN, FIB-SEM, and nano-CT methods. Mar Pet Geol 78:99–109. https://doi.org/10.1016/J.MARPETGEO.2016.09.010

Taylor GH (1966) The electron microscopy of vitrinites. Adv Chem. https://doi.org/10.1021/ba-1966-0055.ch017

Taylor GH, Teichmüller M (1993) Observations on fluorinite and fluorescent vitrinite with the transmission electron microscope. Int J Coal Geol 22:61–82. https://doi.org/10.1016/0166-5162(93)90038-c

Taylor GH, Liu SY, Teichmüller M (1991) Bituminite—a TEM view. Int J Coal Geol 18:71–85. https://doi.org/10.1016/0166-5162(91)90044-j

Taylor GH, Teichmüller M, Davis A, Diessel CFK (1998) Organic petrology

Tiwari P, Deo M, Lin CL, Miller JD (2013) Characterization of oil shale pore structure before and after pyrolysis by using X-ray micro CT. Fuel 107:547–554. https://doi.org/10.1016/J.FUEL.2013.01.006

Trimble AS, Hower JC (2003) Studies of the relationship between coal petrology and grinding properties. Int J Coal Geol 54(3–4):253–260

Valentine BJ, Hackley PC (2019) Applications of correlative light and electron microscopy (CLEM) to organic matter in the north American shale petroleum systems. In: Camp W, Milliken K, Taylor K, Fishman N, Hackley P, Macquaker J (Eds) Mudstone diagenesis: research perspectives for shale hydrocarbon reservoirs, seals, and source rocks, vol 121. AAPG Memoir, pp 1–17

Vishal V, Chandra D (2022) Mechanical response and strain localization in coal under uniaxial loading, using digital volume correlation on X-ray tomography images. Int J Rock Mech Min Sci 154:105103. https://doi.org/10.1016/J.IJRMMS.2022.105103

Vishal V, Chandra D, Bahadur J, Sen D, Hazra B, Mahanta B, Mani D (2019) Interpreting pore dimensions in gas shales using a combination of SEM imaging, small-angle neutron scattering, and low-pressure gas adsorption. Energy Fuels 33:4835–4848. https://doi.org/10.1021/ACS.ENERGYFUELS.9B00442/ASSET/IMAGES/LARGE/EF-2019-00442P_0011.JPEG

Wang Y, Hou ZQ, Hu YZ (2018a) In situ X-ray micro-CT for investigation of damage evolution in black shale under uniaxial compression. Environ Earth Sci 77:1–12. https://doi.org/10.1007/S12665-018-7904-6/FIGURES/12

Wang Y, Li CH, Hao J, Zhou RQ (2018b) X-ray micro-tomography for investigation of meso-structural changes and crack evolution in Longmaxi formation shale during compressive deformation. J Pet Sci Eng 164:278–288. https://doi.org/10.1016/J.PETROL.2018.01.079

Wang G, Jiang C, Shen J, Han D, Qin X (2019) Deformation and water transport behaviors study of heterogenous coal using CT-based 3D simulation. Int J Coal Geol 211:103204. https://doi.org/10.1016/J.COAL.2019.05.011

Wu Y, Misra S, Sondergeld C, Curtis M, Jernigen J (2019) Machine learning for locating organic matter and pores in scanning electron microscopy images of organic-rich shales. Fuel 253:662–676. https://doi.org/10.1016/J.FUEL.2019.05.017

Wu J, Yuan Y, Niu S, Wei X, Yang J (2020) Multiscale characterization of pore structure and connectivity of Wufeng-Longmaxi shale in Sichuan Basin, China. Mar Pet Geol 120:104514. https://doi.org/10.1016/J.MARPETGEO.2020.104514

Xiong Q, Li K, Yang D, Yu H, Pan Z, Song Y (2020) Characterizing coal pore space by gas adsorption, mercury intrusion, FIB–SEM and μ-CT. Environ Earth Sci 79:1–15. https://doi.org/10.1007/S12665-020-08950-3/FIGURES/16

Yang F, Ning Z, Wang Q, Zhang R, Krooss BM (2016) Pore structure characteristics of lower Silurian shales in the southern Sichuan Basin, China: insights to pore development and gas storage mechanism. Int J Coal Geol 156:12–24

Yin H, Zhou J, Jiang Y, Xian X, Liu Q (2016) Physical and structural changes in shale associated with supercritical CO_2 exposure. Fuel 184:289–303. https://doi.org/10.1016/J.FUEL.2016.07.028

Yu H, Zhang Y, Lebedev M, Wang Z, Li X, Squelch A, Verrall M, Iglauer S (2019) X-ray micro-computed tomography and ultrasonic velocity analysis of fractured shale as a function of effective stress. Mar Pet Geol 110:472–482. https://doi.org/10.1016/J.MARPETGEO.2019.07.015

Zhao J, Chen H, Zhang J, Zhang W, Chen G (2022) Quantitative characterization of organic and inorganic pores in shale based on FIB-SEM. Front Earth Sci (lausanne) 10:994312. https://doi.org/10.3389/FEART.2022.994312/BIBTEX

Zhou S, Yan G, Xue H, Guo W, Li X (2016) 2D and 3D nanopore characterization of gas shale in Longmaxi formation based on FIB-SEM. Mar Pet Geol 73:174–180. https://doi.org/10.1016/J.MARPETGEO.2016.02.033

Zhou S, Liu D, Cai Y, Yao Y, Li Z (2017) 3D characterization and quantitative evaluation of pore-fracture networks of two Chinese coals using FIB-SEM tomography. Int J Coal Geol 174:41–54. https://doi.org/10.1016/J.COAL.2017.03.008

Zou G, She J, Peng S, Yin Q, Liu H, Che Y (2020) Two-dimensional SEM image-based analysis of coal porosity and its pore structure. Int J Coal Sci Technol 7:350–361. https://doi.org/10.1007/S40789-020-00301-8/FIGURES/9

Chapter 5
Upscaling for Natural Gas Estimates in Coal and Shale

In a generic sense, the term 'upscaling' means increasing the size or scope of something, usually in a higher order of magnitude than the scale of measurement. Similar to other geological formations, subsurface coal/shale reservoirs are complex and heterogeneous, where most of the geological, geomechanical and hydrological features are scale-dependent and the measurements along with their repeatability strictly depend on the region of interest. For coal and shale reservoirs the natural gas is primarily stored in nanopores hosted in organic matter and clay minerals, hence quantification of the hydrological (porosity, permeability) and geomechanical properties are often performed in core-scale or even finer scales (powders, small splits etc.), due to limitation in experimental techniques. In most cases, the measurements are performed in samples collected from different depths and different boreholes to understand the variation in properties over the area of interest. Abundance of log measurements containing information on porosity, permeability, density, sonic properties etc. can encompass significant depth dependent information, however these log data can only provide information for formations close to the wellbore. Experimentally determined properties along with log data are simplified and used as input parameters in a reservoir simulator, which, in theory populates the information for the entire reservoir, creating a static model. This static model can be used to determine the total gas storage capacity, simulate gas injection or production behavior and forecast the temporal change in reservoir properties after several years of operation.

Field-scale representation of a reservoir is highly challenging and numerically expensive. Although, accurate representation of geological models is crucial to understand the reservoir behavior more accurately; a highly detailed model can increase simulation run time and also hamper dynamic evaluation, which is one of the primary objectives of a numerical simulator. Upscaling in such scenarios is essential, which incorporates approaches like systematic averaging of reservoir properties and minor heterogeneities, which in turn reduces the details of the model, makes the simulations a lot faster with minimal sacrifice in the accuracy of the results obtained. Multiple approaches exist to upscale a reservoir model and the tradeoff between loss

B. Hazra et al., *Unconventional Hydrocarbon Reservoirs: Coal and Shale*, Petroleum Engineering, https://doi.org/10.1007/978-3-031-53484-3_5

of detail and increase in simulation efficiency is an emerging subject for the global research community. History matching the reservoir operation (injection/production data) is an important benchmarking procedure for reservoir models to optimize input parameters. Adaptation of machine learning and artificial neural network in reservoir simulations have also greatly simplified the history matching workflow and enhanced model prediction (Lu et al. 2022; Xu et al. 2020; Yang et al. 2022; Zhu et al. 2022).

5.1 Assumptions in Reservoir Upscaling

Coal reservoirs contain natural fractures termed as cleats, which span in both horizontal and vertical direction with respect to the bedding planes. The dominate sets of cleats are termed as face cleats and the less dominant ones are termed as butt cleats, where the face cleats cuts through the butt cleats. These natural fractures act as conduits for gas/fluid flow and the spacing of these cleats are essential parameters to consider the ease of flow in the reservoirs. Due to presence of these cleat systems, CBM reservoirs are termed as dual-porosity, dual-permeability system, where matrix and cleats represent entities of different porosity and permeability. Development of suitable simulators for CBM applications have been going on since 1970's and the first few iterations were repurposed from conventional black-oil simulators and didn't have both dual-porosity and dual-permeability considerations in them (Cicek 2003; King et al. 1986; Manik et al. 2002; Reeves and Pekot 2001; Remner et al. 1986; Sawyer et al. 1990). Such lack in considerations often resulted in over prediction of gas recovery and underprediction of water recovery. The porosity and permeability models for shale reservoirs are even more complicated. For shales, mineral matrix plays a crucial role alongside organic matrix, therefore they are considered as two different porosity entities. Although the cleat system does not exist in shales, often shale reservoirs are fractured in a controlled way to enhance their permeability and acts as a wider conduit for fluid flow. These fractures are modeled as a separate entity with its own porosity and permeability (Haghshenas et al. 2013; Sheng et al. 2018; Sun et al. 2015; Taghavinejad et al. 2020; Yao et al. 2013). The gases in the matrix of coal and shale are stored in the pores by the process of adsorption (Loucks et al. 2009; Sondergeld et al. 2010; Wang et al. 2009), which is different from the storage mechanism in conventional oil and gas reservoirs. This prompts inclusion of the Langmuir isotherm parameters into the mass balance formulations which can manage pressure dependent influx of gases from matrix to fractures. Flow in fractures and transport of fluids from fractures to wellbores are solved with Darcy equation. Since coal and shale reservoirs are saturated with gas and brine, relative permeability formulation for matrix and fractures are needed (Haghshenas et al. 2013).

5.2 Upscaling in Flow Properties

Laboratory measurements conducted throughout the past few decades have indicated that unconventional reservoir rocks like shale and coal have typically very small pore sizes (in nanometre range) with majority of the pores having a width smaller than 2 nm (Clarkson and Bustin 1999; Loucks et al. 2012; Loucks and Reed 2014; Sun et al. 2015). Gas transport in shale pores and matrix is also a function of pore sizes and their connectivity. At reservoir analog pressure CH_4 gas molecules form a condensate adsorption layer on the surface of pores followed by a mobile and less dense phase transition layer. The thickness of the transition layer is approximately twice the diameter of a methane molecule. Given the very small room for gas transport and the effect of pore walls and throats on the transport mechanism, it very closely resembles Knudsen diffusion mechanism and also prompts to ignore the no slip boundary condition in the continuum flow regime (Clarkson et al. 2011). The nanoscale pores in shale and coal matrix invites complex physics including gas slippage, surface diffusion, ad/de-sorption and renders the conventional Darcy equation inapplicable.

One of the first instances of modeling complex fluid flow in dual porosity medium (fractures and matrix) arrived from Barenblatt's equations (Barenblatt et al. 1960) with an assumption that the fracture system and matrix are in an overlapping continuum. This assumption helped them define the fluid properties at every numerical gridpoint of both domains. In this case, Darcy's law was considered to be valid, and each domain was assumed to be homogeneous and isotropic with no fracture compressibility. Fluid transfer across domains were captured with a source/sink term, which was included in the mass conservation formulation and flow in each domain was determined by the combination of Darcy's law and the conservation equation. Warren and Root (1963) developed the dual porosity model by postulating that the fractured reservoir is a system composed of identical parallelepipeds disturbed by a regularly spaced orthogonal fracture network. No flow boundary condition was assumed between consecutive matrix blocks, but 1D flow was allowed at the fracture-matrix interface with shape factor and transfer function controlling the flow characteristics. Kazemi improved the Warren and Root model and extended it to 3D (Kazemi et al. 1976) while incorporating gravity segregation between each grid blocks. This approach was further extended by incorporating a third phase (Thomas et al. 1983), however none of these models considered capillary pressure in fractures, phase segregation in matrix and operated with an assumption of pseudosteady state flow within matrix blocks. de Swaan (1976) introduced an analytical solution that accounts for the transient flow regime in a dual-porosity model, taking into consideration the unsteady-state flow between the matrix and fractures (Fig. 5.1).

To capture the transient flow and phase segregation in matrix blocks, it was needed to subdivide the matrix into smaller subdomains, however this significantly increased the computational time as more variables needed to be incorporated in each of the subdomains (Beckner et al. 1991; Gilman 1986) and made the simulations prone to failure due to non-convergence of Newton iterations. Contrary to Warren and Root model for flow simulation, some researchers argued that the flow behavior within

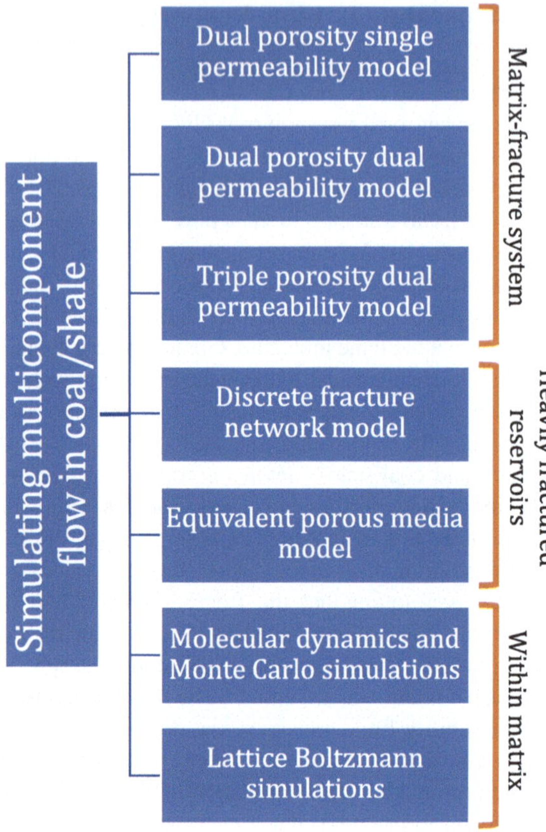

Fig. 5.1 Methods for simulating flow at different scales in fractured coal and shale reservoirs

matrix and across matrix–matrix interface plays a huge role in fluid flow, especially in longer time scale. Earlier approaches ignored this part with an assumption that the matrix permeability is so low that almost no flow will occur in them, and the simulations were only performed in stimulated reservoir volumes (SRV) where natural or hydraulic fractures will boost the flow rate and volume. Discrete fracture network (DFN) model was developed to quantify flow within matrix subdomain (Hofmann et al. 2014; Shiqian et al. 2020; Zhang et al. 2009). In this approach every fracture within the reservoir needs to be uniquely identified and needs to be assigned an individual node to initiate the simulation, which is time consuming if there are dense network of fractures in a reservoir. Dual/triple porosity and dual permeability models which are more common these days for shale gas/CBM simulation are an extended form of classical black oil models (dual porosity single permeability) and are demonstrated to have a better history matching compared to its precursors.

Over the last half-century several models have been suggested of varying complexity to account for the multiple physical processes controlling the flow in the reservoir. Researchers especially focused on resolving flow, diffusion, desorption, and convection in the matrix, which in turn controls the flux between matrix and fracture systems. With multiple phases in the matrix and considering the wettability of the reservoirs, developing a relative permeability model was crucial. The apparent permeability model was introduced which couples viscous flow and Knudsen diffusion (Javadpour 2009) and can provide realistic physics of flow through nanopores, also accounting for the wettability of the media. Since then multiple models were formulated in the last few decades for resolving multiscale complex fluid transport in shales considering pore radius and channel widths, Knudsen number and diffusion coefficient (Dusty-Gas Model) (Akkutlu et al. 2015; Azom and Javadpour 2012; Freeman et al. 2011; Islam and Patzek 2014; Li et al. 2017; Sheng et al. 2018; Singh and Javadpour 2016; Wang et al. 2017; Zhang et al. 2015a, b). The multicomponent multiphysics constitutive equations for gas and water transport in the organic matter, inorganic matter (for shale only) and fracture are as follows (Hall et al. 1994; Sun et al. 2015):

Mass transport for water in inorganic/mineral matter:

$$\frac{\partial \phi_I b_w S_{wI}}{\partial t} = \nabla \cdot \left(k_I \frac{k_{r,wI} b_w}{\mu_w} \nabla \Phi_{wI} \right) + \sigma_I k_I \frac{k_{r,wI} b_w}{\mu_w} \Delta \Phi_{wI-F} \tag{5.1}$$

Mass transport for water in fractures:

$$\frac{\partial \phi_F b_w S_{wF}}{\partial t} = \nabla \cdot \left(k_F \frac{k_{r,wF} b_w}{\mu_w} \nabla \Phi_{wF} \right) + q_{ww} + \sigma_I k_I \frac{k_{r,wI} b_w}{\mu_w} \Delta \Phi_{wF-I} \tag{5.2}$$

Mass transport for gas in organic matter (gas component i)

$$\frac{\partial \rho_{rock} \omega_i}{\partial t} + \frac{\partial \phi_O b_g S_{gO} y_{O,i}}{\partial t} = \sigma_O S_g D_{gO} \Delta C_{gO-I,i} \tag{5.3}$$

Mass transport for gas in minerals (gas component i)

$$\begin{aligned} \frac{\partial \phi_I b_g S_{gI} y_{I,i}}{\partial t} &= \nabla \cdot \left(k_I \frac{y_{I,i} k_{r,gI} b_g}{\mu_g} \nabla \Phi_{gI} + S_g D_{gI} \nabla C_{gI,i} \right) \\ &+ \sigma_O S_g D_{gO} \Delta C_{gI-O,i} \\ &+ \sigma_I \left(k_I \frac{y_{I,i} k_{r,gI} b_g}{\mu_g} \Delta \Phi_{gI-F} + S_g D_{gI} \Delta C_{gI-F,i} \right) \end{aligned} \tag{5.4}$$

Mass transport for gas in fractures (gas component i)

$$\frac{\partial \phi_F b_g S_{gF} y_{F,i}}{\partial t} = \nabla \cdot \left(k_F \frac{y_{F,i} k_{r,gF} b_g}{\mu_g} \nabla \Phi_{gF} \right) + y_{F,i} q_{gw} + \sigma_I \left(k_I \frac{y_{I,i} k_{r,gI} b_g}{\mu_g} \Delta \Phi_{gF-I} + S_g D_{gI} \Delta C_{gF-I,i} \right) \quad (5.5)$$

In Eqs. 5.1–5.5, ϕ_O, ϕ_I and ϕ_F denote the porosity of organic material, inorganic material, and fractures, respectively; S_{gO}, S_{gI} and S_{gF} refer to the gas phase saturation of organic material, inorganic material, and fractures, respectively; $y_{O,i}$, $y_{I,i}$, and $y_{F,i}$ are the mole fraction of component i in the gas phase of organic material, inorganic material, and fractures, respectively; S_{wI}, and S_{wF} refer to the water phase saturation of inorganic material and fractures, respectively; $k_{r,wI}$ and $k_{r,wF}$ refer to the water phase relative permeability of inorganic material and fractures, respectively; $k_{r,gI}$ and $k_{r,gF}$ refer to the gas phase relative permeability of inorganic material and fractures, respectively; ϕ_{wI} and ϕ_{wF} are water phase potential of inorganic material and fractures, respectively; ϕ_{gI} and ϕ_{gF} are gas phase potential of inorganic material and fractures, respectively; k_I and k_F are inorganic material and fracture permeability (md), respectively; μ_w and μ_g are water and gas phase viscosity (cp), respectively; σ_O and σ_I are shape factors (ft^{-2}) of organic and inorganic material, respectively; and q_{ww} and q_{gw} are water and gas well production rate (lbm-mol/ft^3/D), respectively. D_{gO} and D_{gI} are gas diffusivity in organic and inorganic material (ft^2/D), respectively; b_w and b_g are water and gas molar density (lbm-mol/ft^3), respectively; $C_{gO,i}$ $C_{gI,i}$ and $C_{gF,i}$ are gas phase component i concentration (lbm-mol/ft^3) in organic and inorganic material and fractures, respectively; ρ_{rock} is rock density (lbm/ft^3); and ω_i is gas phase component i adsorption amount (lb-mol/lb rock). The entire workflow of this multiphysics multicomponent simulation can be represented as Fig. 5.2.

5.3 Coupled Upscaling of Flow and Mechanics

Understanding the geomechanical properties as well as in-situ stress conditions of shale and coal reservoirs are crucial to understand the hydrocarbon recovery potential of the reservoirs. Shale and coal matrix are quite heterogeneous and contain large number of pores which act as storage sites for the gas. Aside from pores there are numerous microcracks which are scattered across the matrix ranging from nm to um scale. The orientation and density of these fractures are important factors that control how the matrix deforms. Stimulated shale reservoirs contain bigger fracture systems, whereas the mutually perpendicular cleat systems in coal act as wider fractures and affects the geomechanical behavior of the reservoir as a whole. Pore pressure depletion results in increase of the effective stress in the reservoir and pushes the reservoir towards higher stress conditions. Desorption of gases from the matrix also induces strain in the organic and mineral matrix. Depending on the saturation

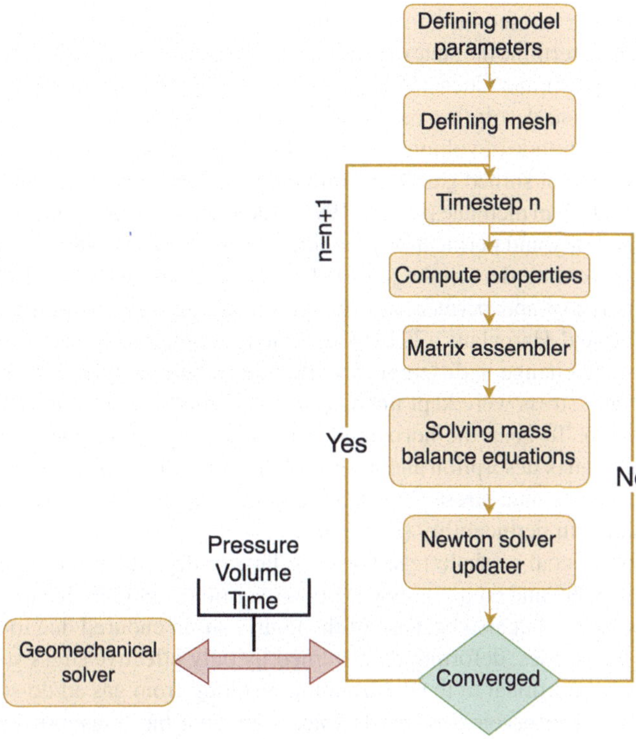

Fig. 5.2 Flowchart of a fully implicit, multicomponent, multiphysics and multiphase flow simulation with a possibility to couple with geomechanics

condition of the reservoirs (saturated, undersaturated and oversaturated), the pressure depletion process varies. For the saturated reservoir, depressurization immediately leads to release of methane from the wellbore, whereas for a undersaturated reservoir, depressurization leads to release of mobile water from the fracture system, followed by release of gases. For an oversaturated reservoir, fracture system is also saturated with mobile gas along with water and depressurization leads to escape of both mobile water and gases from the fractures.

For methane recovery from coal mines, once the reservoir pressure decreases, the flow of gas also becomes unprofitable. In such cases, often CO_2 is injected in the reservoir to drive the flow of methane from the matrix, which is commonly known as ECBM recovery. As discussed before, CO_2 has a higher affinity towards organic matter compared to methane, which leads to release of methane from pore spaces. However, excessive sorption of CO_2 in the pores leads to matrix swelling, which closes the cleats and lowers reservoir permeability. A combination of CO_2 and N_2 as injected phase has shown to reduce the swelling and maintain flow in the reservoir (Vishal 2017a, b; Vishal et al. 2013, 2015).

Several laboratory-based investigations have proven that the effect of gas content in the matrix can determine the state of swelling or shrinkage in coal matrix (Harpalani et al. 1990; Harpalani and Chen 1995; Reznik et al. 1979). Bulk shrinkage volume of coal after methane desorption can be as high as 24.5% based coal rank and the conversely the magnitude of volumetric strain due to gas uptake (swelling) is proportional to the volume of sorbed gas. The strain induced due to swelling can be almost 3× higher for CO_2 than methane (Levine 1996). Laboratory experiments have shown that the cleat porosity and permeability can increase by 80% and 480% respectively with a minimal volumetric shrinkage of 0.16% (Harpalani and Chen 1995). With lower initial porosity and permeability of the cleats, the rate of increase is even higher. Saurabh and Harpalani, (2018) performed a comparative study of desorption induced stress change in different sub-bituminous coal samples. The horizontal strain and vertical stress were kept unchanged while methane from the matrix was allowed to desorb. To keep the horizontal strain unchanged, horizontal stress was reduced during matrix desorption and it showed that all coal samples undergo significant change in horizontal stress (Fig. 5.3a) and accompanied by several orders of magnitude change in permeability (Fig. 5.3b).

Deformation in coal or shale reservoirs are largely dependent on the saturation state of the reservoirs and continuously evolves over time, not only due to the evolution of pore pressure, but also because of the matrix strain endured due to swelling/shrinkage. In most cases, deformation governed by only effective stress state might not be linearly proportional to the deformation resulting from gas ad/de-sorption in matrix. Traditional reservoir flow models accommodate the transport response as an overlapping continuum and neglects the mechanical effects, however for coal/shale where gas storage predominantly relies on sorption, deformation and transport simulation needs to be truly coupled (Fig. 5.4).

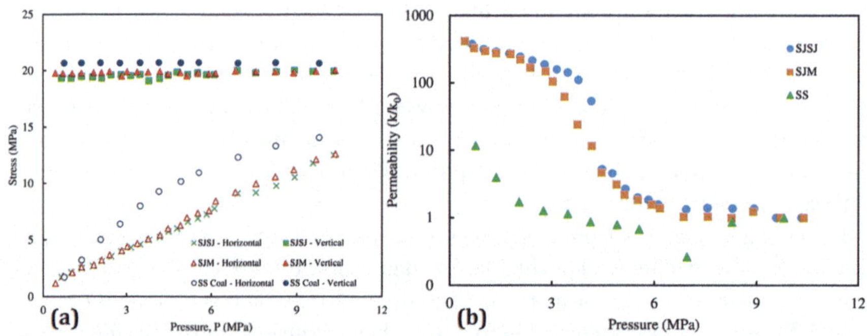

Fig. 5.3 **a** Stress changes endured by coal samples during matrix desorption and **b** corresponding changes in permeability. Reprinted from International Journal of Coal Geology, Vol. 185, Saurabh and Harpalani (2018), Stress path with depletion in coalbed methane reservoirs and stress based permeability modeling, Pages No. 12–22, Copyright (2018), with permission from Elsevier

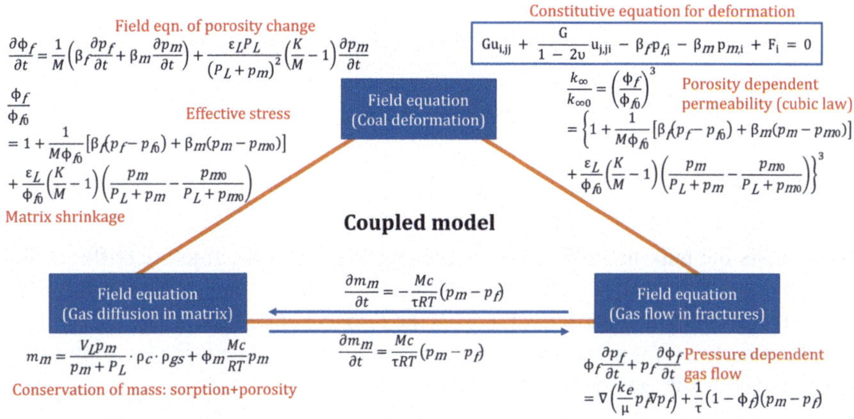

Fig. 5.4 Physics of coupling flow and deformation in coal/shale reservoir simulations [modified after Liu et al. (2015)]

Similar to simulating transport in coal, geomechanical simulation in coal/shale requires a few simplistic approximations. The reservoir is assumed to exist within a a dual poroelastic continuum, where the gases stored in the pores of matrix behave ideally with isothermal boundary conditions. The reservoir is also assumed to be fully saturated with gas to simplify initial conditions. In such configuration, the total strain induced by the reservoir is a summation of strain induced by change in effective stress and by sorption in matrix of the reservoir. The total strain can be represented as follows (Zhao et al. 2020):

$$\varepsilon_{ij} = \frac{1}{2G}\sigma_{ij} - \left(\frac{1}{6G} - \frac{1}{9K}\right)\sigma_{kk}\delta_{ij} + \frac{\alpha p_{\mathrm{m}}}{3K}\delta_{ij} + \frac{\beta p_f}{3K}\delta_{ij} + \frac{\varepsilon_s}{3}\delta_{ij} \tag{5.6}$$

where ε_{ij} is the component of the total strain tensor, G is shear modulus, E is Young's modulus, K is bulk modulus and α and β are Biot's coefficient. These terms can be expressed as follows:

$G = F/2(1+\nu)$, where

$$F = 1/\left[\frac{1}{E} + \frac{1}{aK_n}\right] \tag{5.7}$$

$$K = F/3(1-2\nu) \tag{5.8}$$

$$\alpha = 1 - K/K_s \tag{5.9}$$

$$\beta = 1 - K/K_f \tag{5.10}$$

$$K_f = aK_n \tag{5.11}$$

where K_s is the bulk modulus of individual grains, K_n is the fracture stiffness and K_f is the modified fracture stiffness, ν is Poisson's ratio, p_{m} and p_{f} are gas pressures of matrix and fractures respectively; δ_{ij} is the Kronecker delta function.

Additionally, sorption induced volumetric strain can be fitted into Langmuir type equation, following (Durucan et al. 2009):

$$\varepsilon_s = \varepsilon_L \frac{p_{\mathrm{m}}}{p_{\mathrm{m}} + P_{\mathrm{L}}} \tag{5.12}$$

where ε_L stands for Langmuir volumetric strain, signifying the volumetric strain at infinite pore pressure; P_{L} represents the Langmuir pressure constant, which indicates the pore pressure at which the measured volumetric strain is equal to $\varepsilon_L/2$ (Zhao et al. 2020).

Combining with the equilibrium equation $\sigma_{ij,j} + f_i = 0$ and Cauchy's equation $\varepsilon_{ij} = \left(u_{i,j} + u_{j,i}\right)/2$, the governing equation for shale/coal deformation can be expressed as:

$$Gu_{i,kk} + \frac{G}{1-2\upsilon}u_{k,ki} - \alpha p_{m,i} - \beta p_{f,i} - K\varepsilon_L \frac{P_L}{(p_m + P_L)^2} p_{m,i} + f_i = 0 \tag{5.13}$$

where u_i represents the component of the displacement; σ_{ij} signifies the component of total stress tensor and f_i denotes the component of the body force.

Zhao et al. (2020) simulated the chance in transport properties of a fractured reservoir over time using a coupled geomechanics flow simulation where the porosity and permeability of the matrix and microfractures within the matrix was allowed to change based on dynamic sorption behavior of the matrix (Fig. 5.5). They found that the change in matrix and microcrack permeability with decreasing pressure behaves differently if geomechanical considerations are included in the simulation. With geomechanics, the permeability of the matrix keeps lower at lower pressure compared to the case when geomechanics is not considered. The opposite trend is observed for permeability of the microcracks, which is marginally higher when geomechanics is considered. The combined effect of these two changes also affects the gas production behavior in the reservoir (Fig. 5.5c).

Depending on the method of coupling and cross feedback between deformation and transport properties, these techniques can be classified as loosely coupled, sequential coupled and fully coupled simulations.

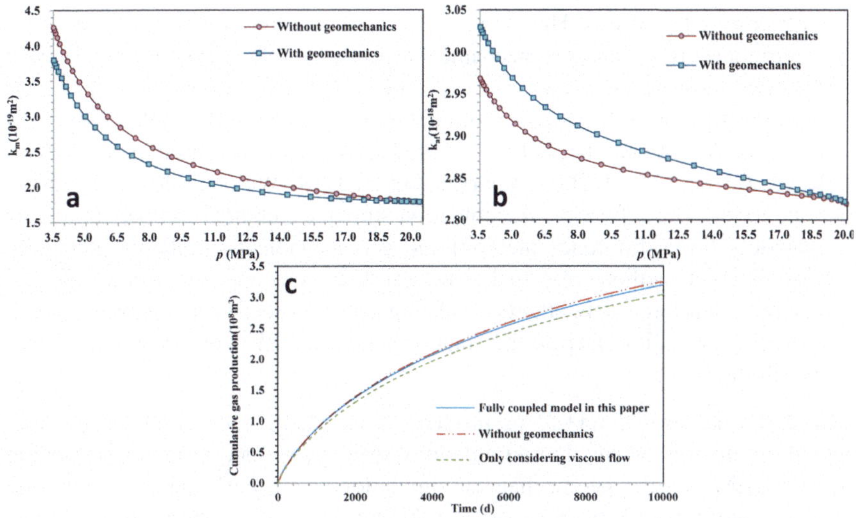

Fig. 5.5 Permeability as a function of pore pressure for **a** matrix and **b** microfractures with and without the consideration of geomechanics. **c** indicates difference in cumulative gas production volume when a coupled model is considered. Reprinted from Journal of Petroleum Science and engineering, Vol. 195, Zhao et al. (2020), Numerical simulation of shale gas reservoirs considering discrete fracture network using a coupled multiple transport mechanisms and geomechanics model, Pages No. 107588, Copyright (2020), with permission from Elsevier

a. **Loosely coupled methods**: For a loosely coupled approach, reservoir flow simulator performs fluid flow calculations independently at each time step and through an intermediate data organizer passes the flow solutions to the geomechanical model, which act as input parameters in geomechanical iterations. This approach is termed loosely coupled since there is only one way flow of information and the outcome of the geomechanical simulations in no way affect the functioning of flow simulations. The frequency of iteration in the flow and geomechanical simulation might also vary. Since the geomechanical results do not affect the flow model, often the mechanics are solved after multiple timesteps in the flow solver.
b. **Sequentially coupled methods**: For a iterative or sequentially coupled approach, both the geomechanical solver and flow solver works separately and sequentially in each time step. The intermediate output from the flow solver is constantly fed into the geomechanical simulator. Output from the geomechanical simulator with the information received from the flow simulator is then passed back onto the flow solver until the total system of equations converge (Kim 2010; Kohl et al. 1997; Rutqvist et al. 2002; Rutqvist and Tsang 2002). In a sequentially coupled simulator, if the coupling parameters are repeated in each time step, then it's termed as an implicit-sequential coupling simulation. Conversely, if the coupling parameters are assumed to be constant at each timestep, then it is called an explicit-sequential coupling.

c. **Fully coupled methods**: This is the best form of coupling, where nonlinear PDFs governing transport and geomechanical aspects are solved concurrently in each time step, however to reduce the order of complexity in competing transport mechanism, most fully coupled simulators only work with single component mass transfer. The number of fully coupled solvers have increased in the past two decade (Minkoff et al. 2003; Noorishad et al. 1984; Wu et al. 2014a, b; Zhang et al. 2016). The difference in simulation strategies adopted in flow simulators (generally finite difference method) and geomechanical simulators (generally finite element method) also makes it hard to truly couple them in a seamless manner. Therefore, more attention is being given on sequential coupling, which with little fine tuning can pass information back and forth between two simulators at each timesteps.

Gu and Chalaturnyk (2005) performed a case study with explicit sequential coupled simulation where flow simulations were performed with a commercial software GEM and the geomechanical simulations were performed with another commercial software FLAC, whereas AutoMate was used to perform the couplings automatically. Their findings indicate that there is a massive difference in gas and water recovery rate when a fully coupled scenario was considered as opposed to a scenario where geomechanics was not considered. As a result, the total gas and water recovery at the end of the simulation period (10 years) increase 1.25× and 3× respectively when geomechanics is included in the simulation. Guo et al. (2018) performed a FEM based coupled flow and geomechanical simulation based on historical production data and reservoir properties of Barnett shale and observed that contrary to Zhao et al. (2020), incorporation of geomechanics while solving flow reduces the flowrate and total recovery of gas (Fig. 5.6a, b). Pressure depletion behavior in the horizontal fractures also appear to be more realistic when geomechanics is incorporated (Fig. 5.6c, d).

Connell (2009) performed a detailed simulation to understand how the gradients in pressure and gas content will vary as a function of distance from the production well. The higher matrix desorption and associated shrinkage near wellbore will cause sinking in the coal reservoir and will vary laterally. They used a combination of SIMED II (for solving flow equations) and FLAC 3D (for solving geomechanics) and iteratively coupled them where at each timestep FLAC 3D performed a quasi-static poroelastic calculation assuming a linear relationship between sorption induced strain and gas content. Although the change in mechanical property due to variation in fluid pressure w is taken into account for the geomechanical calculations, the change in fluid pressure due to geomechanics was not considered for solving the flow equations. Hence, in a way it was a sequentially coupled simulation and not a truly coupled simulation. Their simulation shows that the stress change in the reservoir due to gas production is extremely sensitive to the desorption time of the coal matrix. The vertical stress change is also more significant closer to the production well and less pronounced far from the wellbore. Compared to a purely geomechanical model (Shi-Durucan model (Shi and Durucan 2005), the coupled simulation shows a very high gas production rate in the initial days of production. Shovkun and Espinoza (2017)

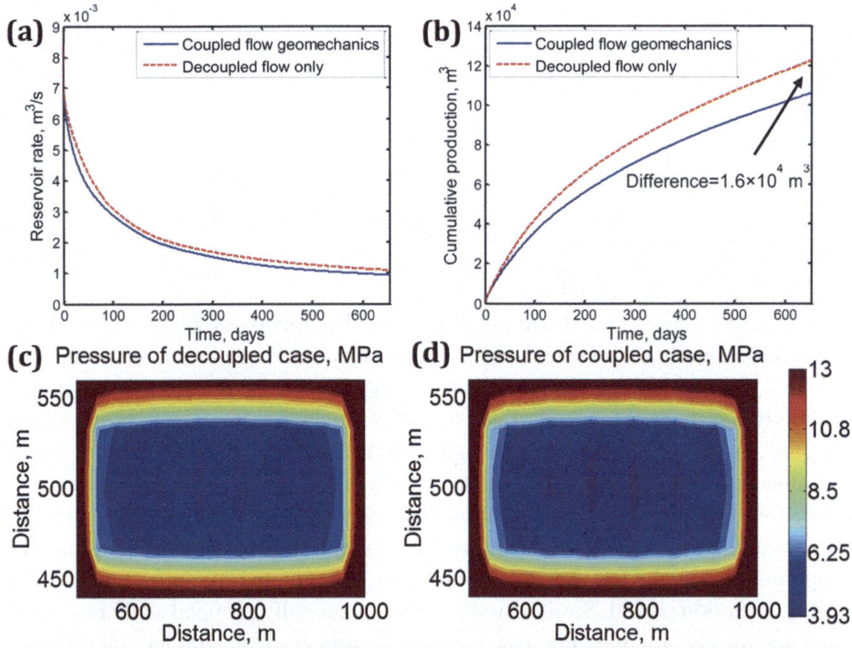

Fig. 5.6 **a** Gas production rate and **b** cumulative gas produced from a fractured shale reservoir considering a coupled approach and uncoupled approach. **c**, **d** represents the pressure field within the fractures considering a decoupled and coupled flow approach. Reprinted from Journal of Petroleum Science and engineering, Vol. 163, Guo et al. (2018), Pressure characteristics and performance of multi-stage fractured horizontal well in shale gas reservoirs with coupled flow and geomechanics, Pages No. 1–15, Copyright (2018), with permission from Elsevier

developed a poromechanical formulation with sorption stress for coal reservoirs which can be expressed as follows:

$$\sigma = \lambda tr(\varepsilon) + 2G\varepsilon + \sigma_s \tag{5.14}$$

And can be further modified into the form of a stress tensor as

$$\begin{bmatrix} \sigma_{xx} \\ \sigma_{yy} \\ \sigma_{xy} \end{bmatrix} = \frac{1}{2G+\lambda}\begin{bmatrix} 4G(G+\lambda) & 2G\lambda & 0 \\ 2G\lambda & 4G(G+\lambda) & 0 \\ 0 & 0 & 2G(2G+\lambda) \end{bmatrix}\begin{bmatrix} \varepsilon_{xx} \\ \varepsilon_{yy} \\ \varepsilon_{xy} \end{bmatrix} + \begin{bmatrix} 1 \\ 1 \\ 0 \end{bmatrix}\left(\frac{S_{zz} - bp}{2G+\lambda} + \sigma_s\right) \tag{5.14}$$

where σ stands for effective stress tensor, ε represents the small strain tensor, λ and G are the Lame's constants of the rock and σ_s represents the swelling stress tensor. Total vertical stress on the reservoir is assumed to be constant in this case. The vertical displacement, principal stress components and strain tensors can be obtained as follows

$$u_z = h\varepsilon_{zz};\begin{cases}\varepsilon_{zz} = \frac{1}{E}\left[\sigma_{zz} - \nu\left(\sigma_{xx} + \sigma_{yy}\right)\right]\\ \varepsilon_{xz} = \frac{1}{2}\left(\frac{\partial u_z}{\partial x} + \frac{\partial u_x}{\partial z}\right)\\ \varepsilon_{yz} = \frac{1}{2}\left(\frac{\partial u_z}{\partial y} + \frac{\partial u_y}{\partial z}\right)\end{cases};\begin{cases}\sigma_{zz} = S_{zz} - bp\\ \sigma_{xz} = 2G\varepsilon_{xz}\\ \sigma_{yz} = 2G\varepsilon_{yz}\end{cases} \quad (5.15)$$

where h is the formation thickness

Their fully coupled model was enabled by the FreeFem++ solver with a Continuous Galerkin method to derive the FEM model. Quadratic and linear basis functions in each of the grid blocks were used to solve the displacement and pore pressure in a fully implicit manner to satisfy Ladyzhenskaya-Babuška-Brezzi conditions. They predicted that the sorption induced change in permeability and porosity is more significant close to the wellbore and accounting for fracture contraction/dilation makes a better fit with the actual field production data for water and gas. They also deduced that although the change in permeability due to sorption mechanism is nominal, the change in horizontal stress due to sorption is significant and might pose hazard for longer production period. Similar studies for hydraulically propped shale reservoirs reveal that the density of natural fractures and their interaction with natural fractures play a pivotal role in determining stress conditions of the reservoir and well performance (Liu et al. 2020). However, the deformation of fractures itself play minor role in the overall stress evolution of the reservoir in the long run, same goes for natural fractures which are disconnected from the hydraulic fractures.

5.4 Role of Machine Learning in Upscaling

Technologies like machine learning (ML) and artificial intelligence (AI) has evolved a lot and became more optimized in the last few decades. These tools have proven to be an undeniable asset in data sciences and has been applied extensively in geological and geophysical data interpretation in the last decade. Introduction of AI and ML in workflow has made realtime predictions and monitoring far more accurate and renders dealing with massive amount of data from logs and optic fiber cables easier. In simple terms, ML enables computers to learn data interpretation through self-defining algorithms based on training data provided. Availability of larger datasets generally make the models perform better, so with little or no human intervention the algorithm will only be better over time. AI and ML techniques not only help interpret larger datasets but can also mend gaps in large datasets by identifying patterns from the rest of the dataset. This ability is very commonly leveraged in filling seismic gaps or gaps in log data based on minimal available information. In the following section we describe a few ML models which has been extensively used in analyzing petrophysics and geomechanics of unconventional reservoirs:

a. **Artificial neural network**: Artificial neural network (ANN) is a scientific ML model consisting of numerous 'neurons' which are arranged in multiple layers depending on the input data and expected output. ANN model consists of three primary layers: input layer, output layer and an intermediate hidden layer, which can have multiple sublayers with complex data mapping based on requirement. This method's primary application is to approximate any function with arbitrary precision. An activation function is utilized to produce outputs in response to inputs. The generated errors in the output are corrected by comparing values to target values, and the algorithm iterates until the generated errors reach a minimum acceptable value (Syed et al. 2022b). ANN models can be used to generate synthetic geomechanical logs based on conventional logs depending on pre-validated correlation data obtained from lab/field scale studies.
b. **Fuzzy inference systems**: Objective of a fuzzy inference system is to determine uncertainty in predicted data and try to analyze the reason of uncertainty. The uncertainties may arise from mismatch of training data and input data or might simply depend on the possibility or probability of the event. Fuzzy system is adept in identifying events when vulnerability arises as a resultant of occasion haphazardness. The system tries several attempts to recognize an association between given input and output data by formulating rules and certain components specified in the system. Compared to conventional logic systems fuzzy logic can provide a probability between true (1) and false (0), which is useful for understanding the interrelationship or consistency of input and output parameters in a quantitative way.
c. **Genetic algorithm:** As the name suggests, genetic algorithm is an approach for addressing both constrained and unconstrained problems, grounded in the principle of natural selection. The algorithm continuously modifies a population of individual solutions. The algorithm generates random population of input data and evaluates them based on an objective function. In each step of generation, a subset of the generated population will be chosen and will be crossed with other members of the subset, which will allow mutation of the datapoints within the subset. A probability of crossover and mutation will be set by the user to constrain the genetic evolution. Once the algorithm finds the best datapoints within the population (threshold set by the user), the algorithm picks these datapoints and restarts crossing them with the untouched members of the population. This process goes on until the threshold set by the user is met.
d. **Support vector machine:** Support Vector Machines (SVMs) are models in supervised learning, well-suited for analyzing data in tasks such as classification and regression. Given a set of training data belonging to one of many classifiers, the algorithm formulates a model that assigns new examples to each of the category making it a non-probabilistic binary linear classifier. This technique is extensively used to determine singular outputs based on a definite mapping of input and output data (Ahmadi and Pournik 2016). Examples include calculation of brittleness index for geomechanical models based on log data.

e. **Decision tree model:** The objective of a decision tree model is to predict the value of a target variable based on multiple input variables. The foundation of this model is a Classification and Regression Tree (CART) model, which iteratively partitions input datasets with the aim of encapsulating the training dataset within the most compact 'tree' possible. The model continuously splits selected variables and prunes excessive branches to minimize complexity. The popularly used Random forest (RF) and Gradient boost algorithms are built on the decision tree model.

Kalantari-Dahaghi et al. (2015) developed a pattern recognition-based proxy model based on ML for Marcellus shale to determine shale gas production at different time resolutions. The simulation results obtained from different clusters of the proxy model on a daily, monthy and yearly basis show $< 15\%$ error when paired with blind simulation runs. They concluded that this kind of scenario can act as an assisted history matching tool, where pattern recognition is more important for figuring out unknown parameters until the best match is obtained. The base case of reservoir model can be fine-tuned with the updated datasets from iterative history matching. Mehana et al. (2021) used a Non-negative Matrix Factorization (NMF) method (Lee and Seung 1999) coupled with k-means clustering to predict the production profile of an existing shale gas well in the Eagle Ford shale play and based on that indicate efficiency of new wells. Vikara et al. (2020) performed a gradient boosted regression tree algorithm to develop a model for accurate prediction of productivity indicators in the Marcellus shale. Their developed model could also provide a fast and efficient indication for well placement and orientation to optimize productivity of the basin. Similar instances of predicting production performace for Eagle Ford shale was adopted by (Syed et al. 2022a), where they used an ANN algorithm coupled with k means clustering to forecast production behavior based on inputs from huge amount of log data. They concluded that inclusion of more data including reservoir completion state, density and orientation of hydraulic fracturing etc. will make the model more robust and quicker in predicting the recovery efficiency. Temizel et al. (2020) used a recurrent neural network model with a decision tree method XGBoost to forecast recovery prediction in shale reservoirs. Song et al. (2021) performed a study to predict geomechanical parameters from well log data of Jimusaer field, west China and used Bidirectional long short-term memory (BiLSTM) algorithm for their purpose. They also indicated that their algorithm produces faster result compared to other decision tree-based algorithms. Similar approach was adopted by (Nath et al. 2022) and they concluded that both random forest and BiLSTM show similar accuracy in predicting geomechanical properties from well log data. Similar work by (Lv et al. 2021; Naeini et al. 2019; Sanei et al. 2023) is worth mentioning here, where physics informed deep learning model or deep neural network models were used to predict geomechanical or flow attributes (permeability) from conventional log data.

5.5 Commercial and Open-Source Simulators

Based on our discussion so far, it is well understood that there are various approaches for upscaling small scale experimental/logging data to reservoir scale models. Depending on the state of coupling the approach of these simulators and their governing equation varies. Most of the flow simulators are based on black oil simulators developed for conventional oil and gas operations with addition of sorption physics. Although there are numerous papers and reports on reservoir flow and geomechanics simulation, there are only a few packaged codes in comparison, which enable other users to adapt the formulation with minimal tuning. A well packaged code (commercial/non-commercial) should come with a well-documented manual and preferably a few example case studies, which the users can play with while learning the simulation.

Most popular simulators are usually commercial codes, which is well documented, regularly updated and fast with a very good support community. Moreover, simulation approaches, including meshing mechanism and governing equations of such commercial codes are also validated with reservoir production data, which automatically inflates the quality of data gathered from such simulators. ECLIPSE (https://www.software.slb.com/products/eclipse), PETREL (https://www.software.slb.com/products/petrel), INTERSECT (https://www.software.slb.com/products/intersect) etc. are a few simulators developed by SLB which can perform reservoir simulations in CBM and shale gas reservoirs. With proper set of data from the reservoir, such simulators can also perform loosely coupled/sequentially coupled geomechanical simulations. CMG-GEM (https://www.cmgl.ca/gem) and COMET3 (Sawyer et al. 1990) are one of the popular commercial softwares dedicatedly used for modeling flow properties in unconventional reservoirs. TOUGH (https://tough.lbl.gov/) is also a powerful software to simulate non-isothermal multiphase flow in porous media with widespread application in conventional oil reservoirs, unconventional reservoirs, gas hydrates and geothermal applications. Given the barebone structure of TOUGH, its relatively easy to couple it with other solvers like FLAC (https://www.itascainternational.com/software/flac3d) which can iteratively solve the geomechanics and communicate back to TOUGH (Rinaldi et al. 2017; Rutqvist 2011; Rutqvist et al. 2011; Sasaki and Rutqvist 2023). SIMED (https://research.csiro.au/ugre/products/simed/) developed by CSIRO, AU and TNO, NL is also a powerful simulator for determining relationships between reservoir properties, well operation and gas production behavior, which was also on several occasions coupled with FLAC to perform coupled flow-mechanics simulations (Connell 2009; Connell and Detournay 2009; Li and Saghafi 2014; Pan and Connell 2009; Wu et al. 2018). Similar approach was also attempted with a combination of flow simulator IPARS and geomechanical simulator JAS3D (Minkoff et al. 2003). For single-phase flow, the majority of THM simulators can address the coupled equations using a fully-coupled method. Examples include ROCMAS (Noorishad and Tsang 1996), COMPASS (Thomas and Sansom 1995), FRACON (Nguyen 1996), CODE_Bright (Olivella et al. 1994) and Open-Geosys (Wang and Kolditz 2007). Apart from these codes, there are some simulators which can also perform fully coupled simulations for multiphase flow as well, however such simulators are limited e.g. DYNAFLOW (Preisig and Prévost 2011), TOUGH-CSM (Zhang et al. 2015a, b), CODE_Bright and COMPASS. Among examples of iterative sequential coupling of flow and mechanics, mechanical GEOSIM (Settari and Walters 2001), Rocflow (Wang and Kolditz 2007) and FRACTure (Kohl et al. 1997) are worth mentioning. Currently reservoir simulations are being evolved with fully coupled approach and reduced order models with incorporation of ML algorithm, GPU accelerated processes and code parallelization, which can solve complex and large datasets with better accuracy and efficiency, merging the gap between simulation-based forecasting and ground truth.

References

Ahmadi MA, Pournik M (2016) A predictive model of chemical flooding for enhanced oil recovery purposes: application of least square support vector machine. Petroleum 2:177–182. https://doi.org/10.1016/J.PETLM.2015.10.002

Akkutlu IY, Efendiev Y, Savatorova V (2015) Multi-scale asymptotic analysis of gas transport in shale matrix. Transp Porous Media 107:235–260. https://doi.org/10.1007/S11242-014-0435-Z/FIGURES/16

Azom PN, Javadpour F (2012) Dual-continuum modeling of shale and tight gas reservoirs. Proceedings—SPE annual technical conference and exhibition, vol 4, pp 2667–2688. https://doi.org/10.2118/159584-MS

Barenblatt GI, Zheltov IP, Kochina IN (1960) Basic concepts in the theory of seepage of homogeneous liquids in fissured rocks [strata]. J Appl Math Mech 24:1286–1303. https://doi.org/10.1016/0021-8928(60)90107-6

Beckner B, Chan H, McDonald A, Wooten S, Jones T, Corp D (1991) Simulating naturally fractured reservoirs using a subdomain method. https://doi.org/10.2118/21241-MS

Cicek O (2003) Compositional and non-isothermal simulation of CO_2 sequestration in naturally fractured reservoirs/coalbeds: development and verification of the model. In: Proceedings—SPE annual technical conference and exhibition, pp 2485–2493. https://doi.org/10.2118/84341-MS

Clarkson CR, Bustin RM (1999) Effect of pore structure and gas pressure upon the transport properties of coal: a laboratory and modeling study. 1. Isotherms and pore volume distributions. Fuel 78:1333–1344. https://doi.org/10.1016/S0016-2361(99)00055-1

Clarkson CR, Jensen JL, Blasingame TA (2011) Reservoir engineering for unconventional gas reservoirs: what do we have to consider? Society of petroleum engineers—SPE Americas unconventional gas conference 2011, UGC 2011, pp 936–980. https://doi.org/10.2118/145080-MS

Connell LD (2009) Coupled flow and geomechanical processes during gas production from coal seams. Int J Coal Geol 79:18–28. https://doi.org/10.1016/J.COAL.2009.03.008

Connell LD, Detournay C (2009) Coupled flow and geomechanical processes during enhanced coal seam methane recovery through CO_2 sequestration. Int J Coal Geol 77:222–233. https://doi.org/10.1016/J.COAL.2008.09.013

de Swaan OA (1976) Analytic solutions for determining naturally fractured reservoir properties by well testing. Soc Pet Eng J 16:117–122. https://doi.org/10.2118/5346-PA

Durucan S, Ahsanb M, Shia JQ (2009) Matrix shrinkage and swelling characteristics of European coals. Energy Procedia 1:3055–3062. https://doi.org/10.1016/J.EGYPRO.2009.02.084

Freeman CM, Moridis GJ, Blasingame TA (2011) A numerical study of microscale flow behavior in tight gas and shale gas reservoir systems. Transp Porous Media 90:253–268. https://doi.org/10.1007/S11242-011-9761-6/METRICS

Gilman JR (1986) An efficient finite-difference method for simulating phase segregation in the matrix blocks in double-porosity reservoirs. SPE Reserv Eng 1:403–413. https://doi.org/10.2118/12271-PA
Gu F, Chalaturnyk RJ (2005) Analysis of coalbed methane production by reservoir and geomechanical coupling simulation. J Can Pet Technol 44:33–42. https://doi.org/10.2118/05-10-03/32270/ANALYSIS-OF-COALBED-METHANE-PRODUCTION-BY
Guo X, Song H, Wu K, Killough J (2018) Pressure characteristics and performance of multi-stage fractured horizontal well in shale gas reservoirs with coupled flow and geomechanics. J Pet Sci Eng 163:1–15. https://doi.org/10.1016/J.PETROL.2017.12.038
Haghshenas B, Clarkson CR, Chen S (2013) Multi-porosity, multi-permeability models for shale gas reservoirs. In: Society of petroleum engineers—SPE Canadian unconventional resources conference 2013—unconventional becoming conventional: lessons learned and new innovations, vol 2, pp 1092–1101. https://doi.org/10.2118/167220-MS
Hall FE, Zhou C, Gasem KAM, Robinson RL, Yee D (1994) Adsorption of pure methane, nitrogen, and carbon dioxide and their binary mixtures on wet Fruitland coal. In: Proceedings—SPE Eastern regional conference and exhibition 329–344. https://doi.org/10.2118/29194-MS
Harpalani S, Chen G (1995) Estimation of changes in fracture porosity of coal with gas emission. Fuel 74:1491–1498. https://doi.org/10.1016/0016-2361(95)00106-F
Harpalani S, Of Arizona U, Schraufnagel RA (1990) Influence of matrix shrinkage and compressibility on gas production from coalbed methane reservoirs. https://doi.org/10.2118/20729-MS
Hofmann H, Babadagli T, Zimmermann G (2014) Numerical simulation of complex fracture network development by hydraulic fracturing in naturally fractured ultratight formations. J Energy Resour Technol, Trans ASME 136. https://doi.org/10.1115/1.4028690/372920
Islam A, Patzek T (2014) Slip in natural gas flow through nanoporous shale reservoirs. J Unconventional Oil Gas Resour 7:49–54. https://doi.org/10.1016/J.JUOGR.2014.05.001
Javadpour F (2009) Nanopores and apparent permeability of gas flow in mudrocks (shales and siltstone). J Can Pet Technol 48:16–21. https://doi.org/10.2118/09-08-16-DA
Kalantari-Dahaghi A, Mohaghegh S, Esmaili S (2015) Coupling numerical simulation and machine learning to model shale gas production at different time resolutions. J Nat Gas Sci Eng 25:380–392. https://doi.org/10.1016/J.JNGSE.2015.04.018
Kazemi H, Merrill LS, Porterfield KL, Zeman PR (1976) Numerical simulation of water-oil flow in naturally fractured reservoirs. Soc Petrol Eng J 16:317–326. https://doi.org/10.2118/5719-PA
Kim J (2010) Sequential methods for coupled geomechanics and multiphase flow. Stanford University
King GR, Ertekin T, Schwerer FC (1986) Numerical simulation of the transient behavior of coal-seam degasification wells. SPE Form Eval 1:165–183. https://doi.org/10.2118/12258-PA
Kohl T, Evans KF, Hopkirk RJ, Jung R, Rybach L (1997) Observation and simulation of non-Darcian flow transients in fractured rock. Water Resour Res 33:407–418. https://doi.org/10.1029/96WR03495
Lee DD, Seung HS (1999) Learning the parts of objects by non-negative matrix factorization. Nature 401:6755 401, 788–791. https://doi.org/10.1038/44565
Levine JR (1996) Model study of the influence of matrix shrinkage on absolute permeability of coal bed reservoirs. Geol Soc Spec Publ 109:197–212. https://doi.org/10.1144/GSL.SP.1996.109.01.14
Li G, Saghafi A (2014) Comparing potentials for gas outburst in a Chinese anthracite and an Australian bituminous coal mine. Int J Min Sci Technol 24:391–396. https://doi.org/10.1016/J.IJMST.2014.03.018
Li Z, Zhang X, Liu Y (2017) Pore-scale simulation of gas diffusion in unsaturated soil aggregates: accuracy of the dusty-gas model and the impact of saturation. Geoderma 303:196–203. https://doi.org/10.1016/J.GEODERMA.2017.05.008
Liu Q, Cheng Y, Zhou H, Guo P, An F, Chen H (2015) A mathematical model of coupled gas flow and coal deformation with gas diffusion and Klinkenberg effects. Rock Mech Rock Eng 48:1163–1180. https://doi.org/10.1007/S00603-014-0594-9/TABLES/5

Liu L, Liu Y, Yao J, Huang Z (2020) Efficient coupled multiphase-flow and geomechanics modeling of well performance and stress evolution in shale-gas reservoirs considering dynamic fracture properties. SPE J 25:1523–1542. https://doi.org/10.2118/200496-PA

Loucks RG, Reed RM, Ruppel SC, Jarvie DM (2009) Morphology, genesis, and distribution of nanometer-scale pores in siliceous mudstones of the Mississippian Barnett shale. J Sediment Res 79:848–861. https://doi.org/10.2110/JSR.2009.092

Loucks RG, Reed RM, Ruppel SC, Hammes U (2012) Spectrum of pore types and networks in mudrocks and a descriptive classification for matrix-related mudrock pores. Am Assoc Pet Geol Bull 96:1071–1098. https://doi.org/10.1306/08171111061

Loucks RG, Reed RM (2014) Scanning-electron-microscope petrographic evidence for distinguishing organic-matter pores associated with depositional organic matter versus migrated organic matter in mudrocks. GCAGS J

Lu C, Zhang S, Xue D, Xiao F, Liu C (2022) Improved estimation of coalbed methane content using the revised estimate of depth and CatBoost algorithm: a case study from southern Sichuan Basin, China. Comput Geosci 158:104973. https://doi.org/10.1016/J.CAGEO.2021.104973

Lv A, Cheng L, Aghighi MA, Masoumi H, Roshan H (2021) A novel workflow based on physics-informed machine learning to determine the permeability profile of fractured coal seams using downhole geophysical logs. Mar Pet Geol 131:105171. https://doi.org/10.1016/J.MARPETGEO.2021.105171

Manik J, Ertekin T, Kohler TE (2002) Development and validation of a compositional coalbed simulator. J Can Pet Technol 41:39–45. https://doi.org/10.2118/02-04-03/2168622/PETSOC-02-04-03.PDF/1

Mehana M, Guiltinan E, Vesselinov V, Middleton R, Hyman JD, Kang Q, Viswanathan H (2021) Machine-learning predictions of the shale wells' performance. J Nat Gas Sci Eng 88:103819. https://doi.org/10.1016/J.JNGSE.2021.103819

Minkoff SE, Stone CM, Bryant S, Peszynska M, Wheeler MF (2003) Coupled fluid flow and geomechanical deformation modeling. J Pet Sci Eng 38:37–56. https://doi.org/10.1016/S0920-4105(03)00021-4

Naeini EZ, Green S, Russell-Hughes I, Rauch-Davies M (2019) An integrated deep learning solution for petrophysics, pore pressure, and geomechanics property prediction. Lead Edge 38:53–59. https://doi.org/10.1190/TLE38010053.1

Nath F, Asish SM, Ganta D, Debi HR, Aguirre G, Aguirre E (2022) Artificial intelligence model in predicting geomechanical properties for shale formation: a field case in Permian Basin. Energies 15:8752. https://doi.org/10.3390/EN15228752

Nguyen TS (1996) Description of the computer code FRACON. Dev Geotech Eng 79:539–544. https://doi.org/10.1016/S0165-1250(96)80043-5

Noorishad J, Tsang CF (1996) ROCMAS simulator; a thermohydromechanical computer code. Dev Geotech Eng 79:551–558. https://doi.org/10.1016/S0165-1250(96)80045-9

Noorishad J, Tsang CF, Witherspoon PA (1984) Coupled thermal-hydraulic-mechanical phenomena in saturated fractured porous rocks: numerical approach. J Geophys Res Solid Earth 89:10365–10373. https://doi.org/10.1029/JB089IB12P10365

Olivella S, Carrera J, Gens A, Alonso EE (1994) Nonisothermal multiphase flow of brine and gas through saline media. Transp Porous Media 15:271–293. https://doi.org/10.1007/BF00613282/METRICS

Pan Z, Connell LD (2009) Comparison of adsorption models in reservoir simulation of enhanced coalbed methane recovery and CO_2 sequestration in coal. Int J Greenhouse Gas Control 3:77–89. https://doi.org/10.1016/J.IJGGC.2008.05.004

Preisig M, Prévost JH (2011) Coupled multi-phase thermo-poromechanical effects. Case study: CO_2 injection at In Salah, Algeria. Int J Greenhouse Gas Control 5:1055–1064. https://doi.org/10.1016/J.IJGGC.2010.12.006

Reeves S, Pekot L (2001) Advanced reservoir modeling in desorption-controlled reservoirs. In: SPE Rocky Mountain regional meeting proceedings. https://doi.org/10.2118/71090-MS

Remner DJ, Ertekin T, Sung W, King GR (1986) A parametric study of the effects of coal seam properties on gas drainage efficiency. SPE Reserv Eng 1:633–646. https://doi.org/10.2118/13366-PA

Reznik AA, Fulton PF, Lien CL (1979) Measurement of the flow properties of coals for in-situ gasification. Final report, December 1977–November 1978. [With dependence on degree of water saturation]. https://doi.org/10.2172/6279657

Rinaldi AP, Rutqvist J, Finsterle S, Liu HH (2017) Inverse modeling of ground surface uplift and pressure with iTOUGH-PEST and TOUGH-FLAC: the case of CO_2 injection at In Salah, Algeria. Comput Geosci 108:98–109. https://doi.org/10.1016/J.CAGEO.2016.10.009

Rutqvist J (2011) Status of the TOUGH-FLAC simulator and recent applications related to coupled fluid flow and crustal deformations. Comput Geosci 37:739–750. https://doi.org/10.1016/J.CAGEO.2010.08.006

Rutqvist J, Tsang CF (2002) A study of caprock hydromechanical changes associated with CO_2-injection into a brine formation. Environ Geol 42:296–305. https://doi.org/10.1007/S00254-001-0499-2/METRICS

Rutqvist J, Wu YS, Tsang CF, Bodvarsson G (2002) A modeling approach for analysis of coupled multiphase fluid flow, heat transfer, and deformation in fractured porous rock. Int J Rock Mech Min Sci 39:429–442. https://doi.org/10.1016/S1365-1609(02)00022-9

Rutqvist J, Ijiri Y, Yamamoto H (2011) Implementation of the Barcelona Basic Model into TOUGH–FLAC for simulations of the geomechanical behavior of unsaturated soils. Comput Geosci 37:751–762. https://doi.org/10.1016/J.CAGEO.2010.10.011

Sanei M, Ramezanzadeh A, Delavar MR (2023) Applied machine learning-based models for predicting the geomechanical parameters using logging data. J Pet Explor Prod Technol 13:2363–2385. https://doi.org/10.1007/S13202-023-01687-2/FIGURES/13

Sasaki T, Rutqvist J (2023) The impact of the anisotropy of shale creep on the long-term stress evolution of a geological nuclear waste repository. https://doi.org/10.56952/ARMA-2023-0945

Saurabh S, Harpalani S (2018) Stress path with depletion in coalbed methane reservoirs and stress based permeability modeling. Int J Coal Geol 185:12–22. https://doi.org/10.1016/J.COAL.2017.11.005

Sawyer WK, Paul GW, Schraufnagel RA (1990) Development and application of a 3-D coalbed simulator. In: Annual technical meeting, ATM 1990 1191–1199. https://doi.org/10.2118/90-119

Settari A, Walters DA (2001) Advances in coupled geomechanical and reservoir modeling with applications to reservoir compaction. SPE J 6:334–342. https://doi.org/10.2118/74142-PA

Sheng G, Javadpour F, Su Y (2018) Effect of microscale compressibility on apparent porosity and permeability in shale gas reservoirs. Int J Heat Mass Transf 120:56–65. https://doi.org/10.1016/J.IJHEATMASSTRANSFER.2017.12.014

Shi JQ, Durucan S (2005) A model for changes in coalbed permeability during primary and enhanced methane recovery. SPE Reservoir Eval Eng 8:291–299. https://doi.org/10.2118/87230-PA

Shiqian X, Yuyao L, Yu Z, Sen W, Qihong F (2020) A history matching framework to characterize fracture network and reservoir properties in tight oil. J Energy Resour Technol, Trans ASME 142. https://doi.org/10.1115/1.4044767/960837

Shovkun I, Espinoza DN (2017) Coupled fluid flow-geomechanics simulation in stress-sensitive coal and shale reservoirs: Impact of desorption-induced stresses, shear failure, and fines migration. Fuel 195:260–272. https://doi.org/10.1016/J.FUEL.2017.01.057

Singh H, Javadpour F (2016) Langmuir slip-Langmuir sorption permeability model of shale. Fuel 164:28–37. https://doi.org/10.1016/J.FUEL.2015.09.073

Sondergeld CH, Ambrose RJ, Energy D, Rai CS, Moncrieff J (2010) Micro-structural studies of gas shales, 23–25. https://doi.org/10.2118/131771-MS

Song L, Liu Z, Li C, Ning C, Hu Y, Wang Y, Hong F, Tang W, Zhuang Y, Zhang R, Zhang Y, Zhang Q, (2021) Prediction and analysis of geomechanical properties of Jimusaer shale using a machine learning approach. https://doi.org/10.30632/SPWLA-2021-0089

Sun H, Chawathé A, Hoteit H, Shi X, Li L (2015) Understanding shale gas flow behavior using numerical simulation. SPE J 20:142–154. https://doi.org/10.2118/167753-PA

Syed FI, Alnaqbi S, Muther T, Dahaghi AK, Negahban S (2022a) Smart shale gas production performance analysis using machine learning applications. Pet Res 7:21–31. https://doi.org/10.1016/J.PTLRS.2021.06.003

Syed FI, AlShamsi A, Dahaghi AK, Neghabhan S (2022b) Application of ML & AI to model petrophysical and geomechanical properties of shale reservoirs—a systematic literature review. Petroleum 8:158–166. https://doi.org/10.1016/J.PETLM.2020.12.001

Taghavinejad A, Sharifi M, Heidaryan E, Liu K, Ostadhassan M (2020) Flow modeling in shale gas reservoirs: a comprehensive review. J Nat Gas Sci Eng 83:103535. https://doi.org/10.1016/J.JNGSE.2020.103535

Temizel C, Canbaz CH, Saracoglu O, Putra D, Baser A, Erfando T, Krishna S, Saputelli L (2020) Production forecasting in shale reservoirs using LSTM method in deep learning. In: Unconventional resources technology conference, Virtual, 20–22 July 2020. https://doi.org/10.15530/URTEC-2020-2878

Thomas HR, Sansom MR (1995) Fully coupled analysis of heat, moisture, and air transfer in unsaturated soil. J Eng Mech 121:392–405. https://doi.org/10.1061/(ASCE)0733-9399(1995)121:3(392)

Thomas LK, Dixon TN, Pierson RG (1983) Fractured reservoir simulation. Soc Petrol Eng J 23:42–54. https://doi.org/10.2118/9305-PA

Vikara D, Remson D, Khanna V (2020) Gaining perspective on unconventional well design choices through play-level application of machine learning modeling. Upstream Oil Gas Technol 4:100007. https://doi.org/10.1016/J.UPSTRE.2020.100007

Vishal V (2017b) Saturation time dependency of liquid and supercritical CO_2 permeability of bituminous coals: implications for carbon storage. Fuel 192:201–207. https://doi.org/10.1016/J.FUEL.2016.12.017

Vishal V, Ranjith PG, Singh TN (2013) CO_2 permeability of Indian bituminous coals: Implications for carbon sequestration. Int J Coal Geol 105:36–47. https://doi.org/10.1016/j.coal.2012.11.003

Vishal V, Singh TN, Ranjith PG (2015) Influence of sorption time in CO_2-ECBM process in Indian coals using coupled numerical simulation. Fuel 139:51–58. https://doi.org/10.1016/J.FUEL.2014.08.009

Vishal V (2017a) In-situ disposal of CO_2: liquid and supercritical CO_2 permeability in coal at multiple down-hole stress conditions. J CO_2 Utilization 17, 235–242. https://doi.org/10.1016/J.JCOU.2016.12.011

Wang W, Kolditz O (2007) Object-oriented finite element analysis of thermo-hydro-mechanical (THM) problems in porous media. Int J Numer Methods Eng 69:162–201. https://doi.org/10.1002/NME.1770

Wang J, Luo H, Liu H, Cao F, Li Z, Sepehrnoori K (2017) An integrative model to simulate gas transport and production coupled with gas adsorption, non-Darcy flow, surface diffusion, and stress dependence in organic-shale reservoirs. SPE J 22:244–264. https://doi.org/10.2118/174996-PA

Wang FP, Reed RM, John A, Katherine G (2009) Pore networks and fluid flow in gas shales. In: Proceedings—SPE annual technical conference and exhibition, vol 3, pp 1550–1557. https://doi.org/10.2118/124253-MS

Warren JE, Root PJ (1963) The behavior of naturally fractured reservoirs. Soc Pet Eng J 3:245–255. https://doi.org/10.2118/426-PA

Wu YS, Li J, Ding DY, Wang C, Di Y (2014b) A generalized framework model for the simulation of gas production in unconventional gas reservoirs. SPE J 19:845–857. https://doi.org/10.2118/163609-PA

Wu Y, Pan Z, Zhang D, Lu Z, Connell LD (2018) Evaluation of gas production from multiple coal seams: a simulation study and economics. Int J Min Sci Technol 28:359–371. https://doi.org/10.1016/J.IJMST.2018.03.008

Wu Y-S, Chen Z, Kazemi H, Yin X, Pruess K, Oldenburg C, Winterfeld P, Zhang R (2014) Simulation of coupled processes of flow, transport, and storage of CO_2 in saline aquifers. Pittsburgh, PA, and Morgantown, WV (United States). https://doi.org/10.2172/1167349

Xu X, Rui X, Fan Y, Yu T, Ju Y (2020) Forecasting of coalbed methane daily production based on T-LSTM neural networks. symmetry 2020, vol 12, p 861. https://doi.org/10.3390/SYM12050861

Yang R, Qin X, Liu W, Huang Z, Shi Y, Pang Z, Zhang Y, Li J, Wang T (2022) A physics-constrained data-driven workflow for predicting coalbed methane well production using artificial neural network. SPE J 27:1531–1552. https://doi.org/10.2118/205903-PA

Yao J, Sun H, Fan D, Wang C, Sun Z (2013) Numerical simulation of gas transport mechanisms in tight shale gas reservoirs. Pet Sci 10:528–537. https://doi.org/10.1007/S12182-013-0304-3/METRICS

Zhang R, Winterfeld PH, Yin X, Xiong Y, Wu YS (2015b) Sequentially coupled THMC model for CO_2 geological sequestration into a 2D heterogeneous saline aquifer. J Nat Gas Sci Eng 27:579–615. https://doi.org/10.1016/J.JNGSE.2015.09.013

Zhang R, Yin X, Winterfeld PH, Wu YS (2016) A fully coupled thermal-hydrological-mechanical-chemical model for CO_2 geological sequestration. J Nat Gas Sci Eng 28:280–304. https://doi.org/10.1016/J.JNGSE.2015.11.037

Zhang X, Du C, Deimbacher F, Crick M, Harikesavanallur A (2009) Sensitivity studies of horizontal wells with hyduaulic fracurtes in shale gas reservoirs. In: Society of petroleum engineers—international petroleum technology conference 2009, IPTC 2009, vol 2, pp 881–889. https://doi.org/10.3997/2214-4609-PDB.151.IPTC13338/CITE/REFWORKS

Zhang M, Yao J, Sun H, Zhao J, Fan D, Huang Z, Wang Y (2015) Triple-continuum modeling of shale gas reservoirs considering the effect of kerogen. J Nat Gas Sci Eng 24:252–263. https://doi.org/10.1016/J.JNGSE.2015.03.032

Zhao Y, Lu G, Zhang L, Wei Y, Guo J, Chang C (2020) Numerical simulation of shale gas reservoirs considering discrete fracture network using a coupled multiple transport mechanisms and geomechanics model. J Pet Sci 195:107588. https://doi.org/10.1016/J.PETROL.2020.107588

Zhu J, Zhao Y, Hu Q, Zhang Y, Shao T, Fan B, Jiang Y, Chen Z, Zhao M (2022) Coalbed methane production model based on random forests optimized by a genetic algorithm. ACS Omega 7:13083–13094. https://doi.org/10.1021/ACSOMEGA.2C00519/ASSET/IMAGES/LARGE/AO2C00519_0013.JPEG

Chapter 6
Potential for CO_2 Sequestration in Coal and Shale

Unconventional hydrocarbon reservoirs, especially shales and coals are marked by complex pore structures, matrix permeability, and thus effective estimation of storage capacity in both coals and shales present several challenges (Vishal et al. 2019). Because of their unique ability to adsorb gases, coals and shales are seen as potential targets for sequestrating CO_2 and reduce the CO_2 emissions in the atmosphere for sustainable development of society (Vishal et al. 2021a; Chandra et al. 2023; Wang et al. 2024). In general, sequestration of CO_2 for enhanced hydrocarbon recovery from unconventional reservoirs have received tremendous interest in recent years, while sequestration in deep coal seams has also been seen as a popular option. Thus, understanding the mechanisms of storage of gas in coal and shale reservoirs is very critical and of paramount significance.

6.1 Gas Storage in Coal

Commercial production of gas, mainly methane, from coal reservoirs is fairly well established in several countries around the globe. The mechanism through which the Coal Bed Methane (CBM) originates and exists within coal beds, is complex and completely different from conventional reservoirs. While the purpose of this chapter is not to deal with the origin of CBM and readers are referred to Moore (2012), but in general, CBM can be biogenic (microbial origin) and thermogenic (formed due to coalification), although the contribution of the later to global CBM reserves is approximately 5 times higher than the contribution of the later. To put it very simply, coal begins generating thermogenic gas when a vitrinite reflectance of 0.5–06% is achieved due to a combination of several factors (Clayton 1998). Obviously, variations exist, depending upon the coal composition, the geothermal history of the basins etc. (Whiticar 1994).

B. Hazra et al., *Unconventional Hydrocarbon Reservoirs: Coal and Shale*, Petroleum Engineering, https://doi.org/10.1007/978-3-031-53484-3_6

While generation of methane in coal is one critical aspect, another very critical and significant aspect of coal reservoirs is the mechanisms through which they retain the gases. Methane in coal reservoirs exist as—(a) free gas within the coal cleats and fractures, (b) adsorbed phase in the coal microporous structures, and (c) absorbed phase where methane is dissolved within the coal molecular structure (Murray 1991; Rice et al. 1993; Ceglarsk-Stefańska and Zarębska 2002; Moore 2012). However, amongst these, methane primarily exists in adsorbed state on the pore surfaces in the coal within the coal beds (Levine et al. 1993; Milewska-Duda et al. 2000). It has been estimated that nearly 90–95% of methane present in a coal reservoir, exists in adsorbed state within the micropores (Larsen 2004; Medek et al. 2006). What this also points toward is that pore surface area of coal, and not the pore volume, is more significant in understanding the gas storage capacity of coal reservoirs (Moore 2012), and this is where unconventional reservoirs viz. coal and shale fundamentally differs from conventional petroleum reservoirs. The presence of these pore structures within coals makes them unique and is not only significant for holding gases, but these also control the development of activated carbon and liquefaction reactions in coal (Radovic et al. 1997). Methane in coal pore surfaces is physically adsorbed, and not chemically, as indicated by the low heat of adsorption, rapid sorption rate with changing pressure and temperature, and the reversibility of the equilibrium between the gas and solid (Crosdale et al. 1998).

Landmark works in the last four decades have revealed variables that control the methane adsorption capacity of coal. In general rank has been observed to be the predominant factor that controls the methane adsorption capacity of coals (Kim 1977; Ryan 1992). Similarly, the effect of coal composition on adsorption capacity of methane has also been observed to be profound. Crosdale et al. (1998) observed higher methane adsorption capacity in vitrinite rich bright coals, than in inertinite rich dull coals. Other studies have also highlighted that coals rich in vitrinite exhibit elevated methane adsorption capacity compared to inertinite rich coals over wide coal ranks (Crosdale and Beamish 1993; Lamberson and Bustin 1993). Bustin and Clarkson (1998) for iso rank coals, observed good correlation between vitrinite content, micropore volume determined from experiments of low pressure gas adsorption using CO_2 and high-pressure gas adsorption using methane. Mastalerz et al. (2004) for Indiana coals observed rising sorption capacities of CO_2 with increasing content of vitrinite, especially telocollinite, leading them to suggest that telocollinite has the maximum effect on CO_2 sorption capacity amongst all macerals owing to their favourable pore structure for CO_2 sorption. In their study they also observed that coals with higher liptinite content to have lowest sorption capacity.

However, it is over simplistic to consider that coal petrographic composition and rank can directly control their methane adsorption or gas storage capacity. Several studies have revealed that a second order polynomial relationship exists between rank of coal and methane sorption capacity, with the minimum sorption capacity being shown at the high volatile A bituminous coal rank (Levine et al. 1993; Levy et al. 1997). Bustin and Clarkson (1998) noted no discernible trends between methane storage capacity and coal ranks from multiple basins. Similarly, Levy et al. (1997) showed nearly 50% more methane adsorption capacity of Australian coals than US coals, although having similar rank and composition. Laxminarayana and Crosdale (1999) also for Bowen Basin coals, from Australia, also observed a systematic decrease in gas storage capacity from the rank of high volatile bituminous A to medium volatile bituminous. They correctly inferred that this decrease can be attributed to 'plugging' of the microporous structures in medium volatile coals by hydrocarbons. They also commented that surface area of coals also follows a similar pattern and with debituminization and expulsion of hydrocarbons from coal reservoirs, the accessible micropore systems reopen and allows increased adsorption to take place.

Ettinger et al. (1966) found lower to medium rank inertinite-rich coals to have greater adsorption capacity than coals rich in vitrinite. Faiz et al. (1992, 2007) on the other hand reported that amongst the inertinite group of macerals, different macerals may have different pore structures and consequently different sorption capacities. Mukherjee and Misra (2018) in their review claimed that semifusinites are largely microporous and result in higher sorption capacity while fusinites are macroporous and show lesser sorption capacity. While macroporosity in fusinites are evident in optical microscopes (as these are marked by larger cell lumens), the inferences on microporosity of semifusinites seems doubtful, especially when specific works exist substantiating and highlighting the more abundant microporous character and higher internal surface area of vitrinite-rich coals in comparison to inertinite-rich coals at a specified rank (Thomas and Damberger 1976; Unsworth et al. 1989; Faiz et al. 1992; Levine et al. 1993; Laxminarayana and Crosdale 1999; Mastalerz et al. 2004). In this context is it important to understand that different macerals have different pore structures, and when this is superimposed with the coal rank and the volume and nature of mineral matter, simplistic prediction of methane sorption capacity of coal may be challenging. Another significant reason controlling the gas storage or adsorption capacity in coals is their ash content. Mineral matter content in coals have been observed to be simple diluents, and reduces the gas adsorption capacity of coals (Laxminarayana and Crosdale 1999). Rice et al. (1993) demonstrated that mineral matter in coal has limited capacity to adsorb gases, and consequently its presence in greater quantities in a coal limits the space of the organic matter and thereby reduces the methane sorption capacity. Dutta et al. (2011) through their work on Indian coals pointed that for an increase of 1% in mineral matter, sorption capacity for methane dropped by 1 ml/g. However, the role of clay minerals in coal can also impact the methane storage capacity, and this specially makes sense when we consider the storage capacity of shales (with high clay content) in context of coal reservoirs. Varma et al. (2015) though observed overall negative correlation between Langmuir Volume and ash content of coal, however, they observed a position relation between the TOC-normalised Langmuir Volume and ash content for a set of Permian coals from India. The role of clay minerals specifically (also discussed under Sect. 6.2) can be significant, as they provide surface areas and storage sites for methane and CO_2 (Aringhieri 2004; Cheng and Huang 2004).

Moisture content of coal is also known to critically impact the storage capacity of coal. Moisture in coals, exhibit high sensitivity to the advancement of coal rank, especially at the initial stages of coalification (Moore 2012). Moisture covers a certain portion of the coal surface area and limits the adsorption of gases into the micropore structure, and is hence considered to negatively impact the gas adsorption capacity of coal (Yee et al. 1993; Levy et al. 1997; Crosdale et al. 1998). Krooss et al. (2002a, b) observed a 20–25% drop in gas adsorption capacity in moisture equilibrated coals in comparison to dry coals, from Netherlands. Generally, coal moisture content tends to decrease with an increase in both depth and rank (Sivek et al. 2010). It is thus considered that, with increasing rank, there is less moisture competing with methane for the adsorbing sites, and consequently at higher ranks, greater storage capacity is seen (Moore 2012). However, extensive studies on Australian and US coals (Levy et al. 1997; Joubert et al. 1974), reveal a linear decline in methane sorption capacity as moisture content rises up to a critical threshold, and beyond that increased moisture content in the coal has no impact on the adsorption capacity of the coal as it exists in mobile phase. Day et al. (2008) noted the critical moisture content in coals to be higher for CO_2, than CH_4. Even further, Day et al. (2008) and Weniger et al. (2012) documented higher degree in reduction of CH_4 adsorption capacity (~24%) than CO_2 adsorption capacity (~12%) with increasing moisture content. What this essentially implies is while moisture impact the gas storage capacity, depending upon the gas being adsorbed, variations in adsorption capacity can arise due to the competition between water and gas molecules for particular sorption sites and selectivity between CH_4 and CO_2 sorption (Busch et al. 2006; Busch and Gensterblum 2011; Weniger et al. 2012), i.e., competitive adsorption between CO_2 and CH_4, which is extremely critical when considering enhanced coal bed methane extraction and simultaneous sequestration of CO_2 in coal and shale reservoirs (discussed later). As CO_2 kinetic diameter is smaller than CH_4, the former has greater access to several pore structures (blocked or isolated) which are in accessible to the later (Day et al. 2008).

Similar to moisture content, another critical factor not actually dependent on the coal properties, but on the subsurface conditions or laboratory analytical settings, and that can significantly affect the capacity of gas storage in coals is temperature. Yee et al. (1993) through their research documented that temperature is detrimental to the adsorption capacity of coals, as increasing temperatures causes the gases to be in free state, instead of adsorbed state. Similar findings were reported by Levy et al. (1997), Clarkson and Bustin (1996). Perera et al. (2013) explained this to be caused due to temperature induced increase in kinetic energy and diffusion, leading to breakage of bonds between the adsorbed phase and the coal matrix, resulting in reduction of adsorption capacity. A comprehensive study on the influence of temperature on gas adsorption capacity of coals using multiple gases (viz. CO_2, CH_4, N_2) was conducted by Sakurovs et al. (2007), and documented a strong decrease in adsorption capacity for all gases with increasing temperatures.

Generally, it is accepted that coal porosity framework comprises of larger fractures and smaller pores (Clarkson and Bustin 1999; Harpalani and Chen 1997), and within the smaller pore systems, three size classes exist viz. macropores (>50 nm diameter), mesopores (2–50 nm diameter) and micropores (<2 nm diameter) (Gan et al. 1972; Gregg and Sing 1982). As mentioned previously, gas in coal reservoirs primarily is stored in adsorbed form in the pore surfaces, and several researches have documented that gas is primarily stored within the micropore networks. It is commonly accepted that pores marked by smaller sizes impart highest surface areas, and thereby show the highest gas adsorption capacity (Safaei-Farouji et al. 2023). Yao et al. (2008) reported that pores in coal reservoirs follow a fractal distribution caused by different surface irregularities and heterogeneity, and this heterogeneity can significantly impact the capacity of gas adsorption of a coal reservoir. Consequently, fractal theory has been applied for coal reservoirs, to explain their pore properties and gas storage capacity (Yao et al. 2008; Mahamud and Novo 2008; Sun et al. 2015; Li et al. 2017). As universally established, fractal dimensions (D) vary between 2 and 3 and represents smooth surfaces to irregular, complex surfaces, respectively. Amongst different techniques that are commonly employed for determining fractal dimensions, low pressure N_2 gas adsorption technique, employing the Frenkel-Halsey-Hill (FHH) model has been frequently and convincingly used to calculate fractal dimensions in shales and coals (Yao et al. 2008). Fractal dimensions in coal has been further subdivided into two parts—(a) D1, which corresponds to the finer mesopore filling in the relative pressure range of 0.1–0.5 and is strongly influenced by the Van der Waals force, and (b) D2, which corresponds to the coarser mesopore range that gets saturated at the higher relative pressures of 0.5–1.0 and is driven by capillary condensation. Attempts have been made by several to correlate fractal dimensions (D1 and D2) in coal to the gas adsorption capacity of coal.

One of the most intriguing aspects while measuring pore structures and pore size distributions of coals using low pressure N_2 gas adsorption technique is the low surface areas reported by several researchers. It has been documented by several that when N_2 at -196 °C is used, the specific surface area (SSA) of coals is largely under measured, and only the external surface area is measured, as N_2 fails to enter the complex micropores of coals at such low experimental temperatures, and also possibly due to pore shrinkage (Mahajan 1984; Walker et al. 1988). Gan et al. (1972) observed very low N_2 SSA of coals (≤ 1 m^2/g), with limited number of coals (carbon content varying between 75.5 and 81.5%) showing SSA > 10 m^2/g. Even further, several other works (viz. Parkash and Chakrabartty 1986; Mahajan 1984, 1991; Okolo et al. 2015), show significantly lower N_2 gas adsorption derived SSA of coals, compared to CO_2 gas adsorption derived SSA of coals. Figure 6.1 plots BET SSA calculated from N_2 adsorption versus TOC for coals and shales from India and Turkey (Hazra et al. 2020). The figure clearly presents N_2 derived SSA of coals (high TOC end) is concentrated in a narrow band, and SSA of the samples increases with decreasing TOC content. What this effectively indicates is that coals have lower N_2 surface areas, and is not representative of its microporous character. On the other hand, with decreasing TOC content, the SSA is observed to increase. Hazra et al. (2020) using the data on shales and coals from multiple countries documented the limitations of N_2 in accessing complex organic matter pore structures in coals and also shales, and opined that the reduced N_2 derived SSA for rocks with TOC > 10 wt.% i.e., from carbonaceous shales to coals could be due to incomplete access to the micropores of organic matter, while the higher surface areas in rocks with < 10 wt.%

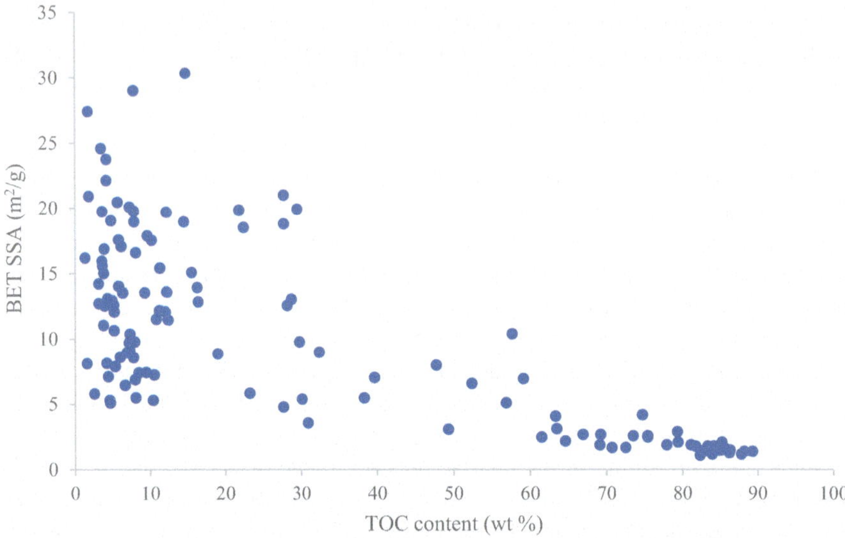

Fig. 6.1 Cross-plot illustrating the correlation between BET SSA and TOC of coals and shales from Turkey and India [data compiled from Mishra et al. (2018), Bakshi et al. (2018), Karayiğit et al. (2018), Hazra et al. (2018a, 2018b, 2019a, 2019b). Reprinted (adapted) with permission from Hazra et al. (2020); Energy Fuels 2021, 35, 1, 456–464. Copyright © 2020 American Chemical Society

TOC could be due to increased access of N_2 to pore structures in the mineral matter rich fractions. In a way, the observations of Hazra et al. (2020) resonate the findings of O'Gorman (1971) who reported < 1 m^2/g SSA of American coals, whereas the ash produced from the inorganic matter through ashing at low-temperature resulted in surface areas of about 10 m^2/g.

In comparison to N_2, CO_2 on the other hand, has been used more convincingly as a potential adsorbate for true representation of microporous structures in coals and shales (Hazra et al. 2020). As CO_2 is marked by smaller molecular dimension compared to N_2, and the adsorption experiments can be conducted at higher experimental temperatures, and hence has been argued to measure the actual and total surface areas of coals (Gan et al. 1972; Walker et al. 1988). Figure 6.2 plots a comparative plot N_2-SSA and CO_2-SSA of coals and shales from different countries. It can be observed that towards the lower ash portion i.e., for coals, the SSA derived using low pressure CO_2 gas adsorption is several times higher than that derived for N_2, whereas towards the higher ash portion i.e., for shales, the SSA from the two adsorbates nearly overlaps. In a pioneering work, Lamond (1962) documented that surface areas measured using N_2 as adsorbate at −196 °C is alike to surface areas measured using CO_2 as adsorbate at −78 °C, for molecular sieves having dimensions of 5 Å, whereas the former adsorbate showed several magnitude SSA than the later adsorbate when molecular sieves with dimensions of 4 Å was used. This led Hazra et al. (2020) to conclude that low pressure gas adsorption experiments using N_2 may

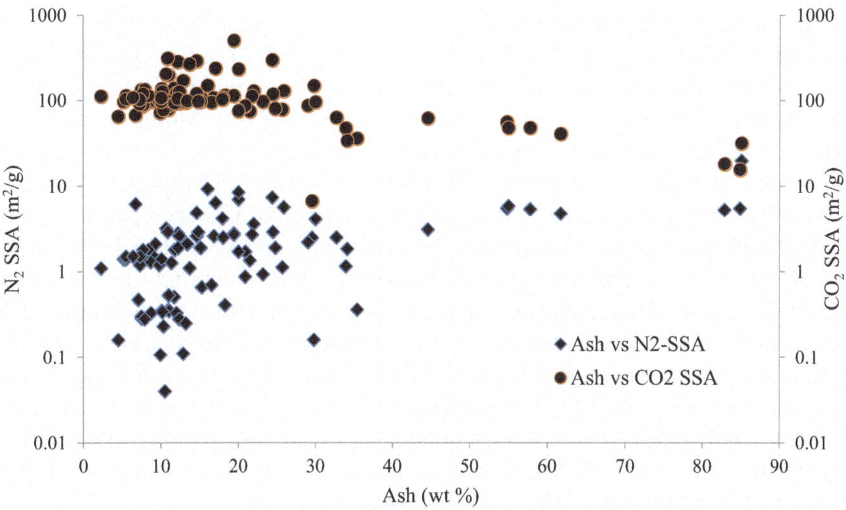

Fig. 6.2 Cross-plot depicting the distribution of SSA determined through N_2 and CO_2 in relation to the ash yield of coals and shales from diverse geographical regions of the world [data compiled from—Parkash and Chakrabartty (1986), Mastalerz et al. (2009), Okolo et al. (2015), Zhao et al. (2016a, b), Chen et al. (2017), Karayiğit et al. (2018), Hazra et al. (2019b). Reprinted (adapted) with permission from Hazra et al. (2020); Energy Fuels 2021, 35, 1, 456–464. Copyright © 2020 American Chemical Society

present inaccuracies in mapping finer pore structures in organic-rich shales (TOC > 10 wt.%) and coals due to molecular sieving effects. However, some researchers have also expressed their concerns and criticisms regarding the outcomes from low pressure CO_2 gas adsorption of coals, mainly due to imbition of CO_2 within the coaly matter, inducing swelling effects, and its interaction with oxygen-containing functional groups of coaly matter (Reucroft and Patel 1983; Mahajan 1991). However, some others have also documented that for pressures at which surface areas of coals are measured, there's limited or no swelling and volume expansion of coals.

Experimental data generated from high pressure gas adsorption capacity of coals employing different gases in the last few decades have showed that CO_2 adsorption capacity of coal consistently remains higher than the CH_4 adsorption capacity (Arri et al. 1992; Greaves et al. 1993). The increased and preferential adsorption of CO_2 relative to CH_4 has made coal reservoirs as excellent targets for sequestration of CO_2, and enhanced Coal Bed Methane (ECBM) recovery whereby it strongly adsorbed to the coal and displaces the CH_4 in its place and thereby accelerating CH_4 production from CBM wells (Schoeling and McGovern 2000; Krooss et al. 2002a, b; Ceglarsk-Stefańska and Zarębska 2002; Harpalani et al. 2006; Day et al. 2008; Weishauptova et al. 2015). Whie initially, the adsorption ratio of CO_2:CH_4 in coal was believed to be 2:1 (Smith 1999; Gentzis 2000), however, subsequent researches have demonstrated that the ratio can vary from > 10 to < 2 in lower ranked coals and medium to low volatile coals, respectively (Stanton et al. 2001b). Because of its lower molecular kinetic diameter (0.33 nm) compared to CH_4 (0.38 nm), and narrower half-width of an exclusive pore (0.289 nm for CO_2 and 0.310 nm for CH_4), CO_2 has increased access to several micropores and ultra-micropores present within coals (Cui et al. 2004; Kelemen and Kwiatek 2009; Safaei-Farouji et al. 2023). Moreover, CO_2 has a greater boiling point (−78.5 °C) relative to methane (−161.5 °C), which causes preferential adsorption of CO_2 over CH_4 during competitive adsorption (Harpalani et al. 2006). Additionally, the larger adsorption energy of CO_2 compared to CH_4 allows it to diffuse into the microporous pore structural framework of coal (Mastalerz et al. 2004). Milewska-Duda et al. (2000) through their modelling study demonstrated that for CO_2, in addition to adsorption by the micropores, a substantial fraction also undergoes absorption into the solid and elastic organic structure of coal, while for CH_4 the amount of adsorption is much lower. All these factors together contribute to the preferential adsorption of CO_2 compared to CH_4 in coal reservoirs, and consequently has been employed as a means to produce extra CH_4 from coal reservoirs, and permanent sequestration of CO_2 in coal reservoirs. Figure 6.3, represents a schematic of competitive adsorption of CO_2 over CH_4 in coal reservoirs.

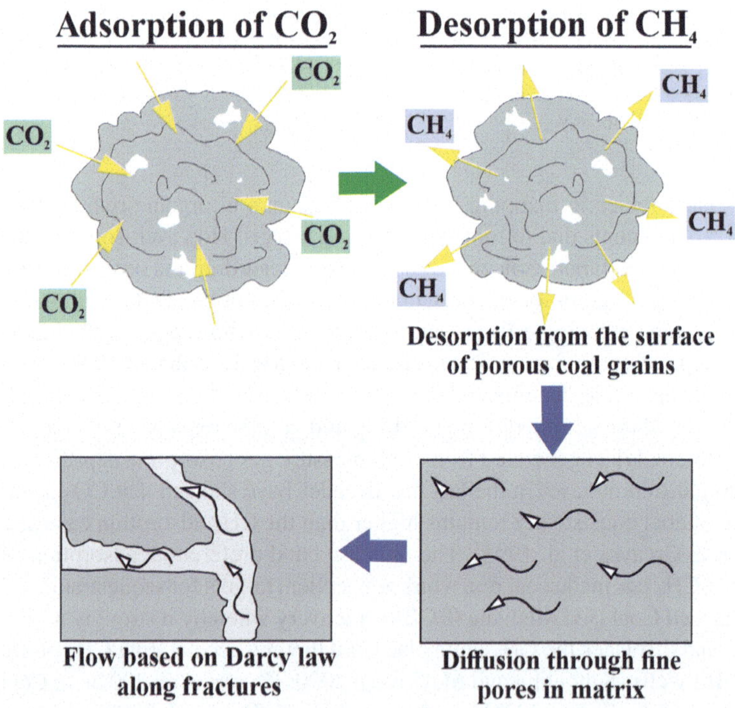

Fig. 6.3 Schematic depicting micro-scale mechanisms during the adsorption of CO_2 and desorption of CH_4 [modified after Harpalani and Schraufnagel (1990)]

During production of CH_4 from coal reservoirs, and subsequent CO_2 injection in coal seams for sequestration and enhanced recovery of CH_4, a number of factors and combination of processes come into play and are considered for achieving the success (Mukherjee and Misra 2018). Infact the CBM industry became effective only when it was realized that CH_4 can be produced profitably from coal reservoirs if the seams are effectively dewatered and by creating a permeable path to the wellbore by means of hydraulic fracturing (Rogers 1994). It is fairly well established that water from coal seams needs to be removed for lowering the pressure and initiating the desorption of CH_4 from the coal seams (Rogers 1994). Desorption of CH_4 is also accompanied by coal-matrix shrinkage, which causes increase in the cleat spacing thereby enhancing the reservoir permeability and reducing the resistance to the flow (Rogers 1994). This desorption induced matrix shrinkage is more prolific towards the end of the well life or at the lower pressure ranges of the adsorption isotherm, where typically large volumes of CH_4 is desorbed with decreasing pressure, until the formation reaches the abandonment pressure point. Overall, during the process of CH_4 production from CBM wells, subsequent CO_2 injection/sequestration and ECBM, different factors, processes viz. effective stresses, Klinkenberg effect, pore pressure, injection pressure etc., are effective, and readers are referred to Rogers (1994) for a detailed understanding. When CO_2 is injected to the coal formation, as mentioned previously, the organic matter preferentially adsorbs (and partly absorbs) the CO_2 and CH_4 is further desorbed in exchange. Opposite to desorption, when the new intrusive gas is adsorbed onto the coal micropores, the coal-matrix swells, jest results in decrement of the cleat-spacings and further drop in permeability. Less surprisingly, adsorption induced matrix swelling will be more prolific for CO_2, than CH_4 (Rogers 1994).

6.2 Gas Storage in Shale

It is projected that natural gas from unconventional shale petroleum systems by 2040 would contribute nearly 30–50% of the total supply of the natural gas in the world (EIA 2016). This scenario of production of gas from unconventional shales, which has to a large extent altered the global energy landscape, has been largely possible due to horizontal drilling and large scale hydraulic fracturing of shale reservoirs, which allows the gas in them to be produced, inspite of their extremely heterogeneous nature and low to ultralow permeabilities (Kerr 2010; Schnoor 2012; Mastalerz et al. 2012, 2013). Moreover, owing to their vast geological reserves and much lower emission index in comparison to burning coal and oil, gas from shale has emerged as vital source of fuel for global energy supply (Curtis 2002; Jarvie et al. 2003, 2007). Similar to CBM reservoirs, gas or oil in unconventional shale reservoirs are generated within the shales and are primarily stored through adsorption (along with being present in free and dissolved states) (Nuttal et al. 2005). Additionally, shales have also received tremendous interest as potential targets for CO_2 sequestration (Tao et al. 2014; Godec et al. 2013; Heller and Zoback 2014; Ma et al. 2018).

Shales are extremely complex and heterogeneous porous materials containing both mineral matter (inorganic) and organic matter (carbonaceous material). Both these phases are heterogeneous and contributes towards a highly complex and heterogeneous pore system in shales (Chalmers et al. 2012; Kuila and Prasad 2013; Milliken et al. 2013; Chandra et al. 2020; Kumar et al. 2024). The overall porosity in shales can be grouped into three types—(i) pores existing within the carbonaceous organic matter; (ii) interparticle: pores located amidst mineral grains; and (iii) intraparticle: pores found within individual mineral grains (Loucks et al. 2009, 2012). The kerogen type/quality, quantity/amount and its thermal maturity level, inorganic/mineral matter, different pore attributes, depth at which the shale horizon is located, and bed thickness, are recognized as the important components from assessing producibility of hydrocarbon and storability of CO_2 in shales (Jarvie et al. 2007; Heller and Zoback 2014; Chalmers and Bustin 2007, 2008; Loucks et al. 2009). For classifying the pore sizes, the classification of the International Union of Pure and Applied Chemistry (IUPAC) is followed, which is as follows: micropores (< 2.0 nm in diameter), mesopores (2.0–50.0 nm in diameter), and macropores (> 50.0 nm in diameter) (Rouquerol et al. 1994; Chalmers and Bustin 2007; Chalmers et al. 2009; Sondergeld et al. 2010). Although CBM and shale gas systems are categorized under unconventional reservoirs and the nature of gas retention is similar in the two reservoirs, a direct comparison of CH_4 and CO_2 adsorption capacities of shales and coals may not be descriptive mainly due to differences in—(a) nature of organic matter (coals are predominantly rich in vitrinites and inertinites), (b) total content of organic matter, and (c) total content of mineral matter and its impact on gas storage (Ross and Bustin 2009).

Over the last 3 decades, numerous authors have carried out noteworthy laboratory investigations to decipher the factors affecting storage capacity of shales. In order to elucidate the influence of minerals in gas storage capacity of shales, Schettler et al. (1991) used pure clays, quartz, and Devonian shales, and observed notable adsorption capacities of minerals. In another significant work on several Devonian shales and pure illite samples, Lu et al. (1995) although observed strong role of TOC in controlling gas adsorption capacity of shales, they also noted illite to play significant role on the same especially for organic-lean samples. In line with these studies, Ross and Bustin (2009) also noted noteworthy microporosity of clay minerals among the irregularly shaped surfaces of clay platelets and also likely due to clay crystal size, enabling them to store gases. Kuila and Prasad (2013) also stressed the importance of mineralogy in controlling the pore-size distributions for shale samples. Importantly, they observed fine mesoporous structures (obtained from low pressure adsorption of N_2 gas) to be linked with illite–smectite clays, while observing none for kaolinites. In another study, Ji et al. (2012) noted highest methane sorption capacity of montmorillonite, compared to illite–smectite mixed layer clays, followed by kaolinite, illite and chlorite, with the latter showing the least gas adsorption capacity. Fan et al. (2014) in their research noted higher methane adsorption capacity of montmorillonite compared to other clay minerals viz. kaolinite, chlorite, and illite, although the adsorption capacity of clays were observed to be much lower than kerogen. All these studies highlight the importance of clay minerals in controlling the gas storage capacity of shales.

Although the effect of clay minerals on gas storage capacity of shales is significant, but the role of organic matter i.e., the carbonaceous matter in shales, has been observed to be most significant towards shales' gas storage capacity (Thomas 2023; Wood and Hazra 2017). Several researchers have established the significance and controls of organic matter richness, type and thermal maturity on gas storage capacity of shales, and of special interest has been the organic matter associated porosity of unconventional shale plays. Different advanced methods such as imaging, gas adsorption, x-ray and neutron scattering have been used to understand and evaluate shale porosity (Mastalerz et al. 2012, 2013; Curtis et al. 2012a, b; Milliken et al. 2013; Labani et al. 2013; Holmes et al. 2017; Hazra et al. 2018a, b; Vishal et al. 2019; Chandra et al. 2022, 2023). Tan et al. (2014) observed sorption capacities of shales of similar thermal maturity levels and clay contents to be positively correlated with their TOC content. Several others also observed strong positive controls of TOC on CH_4 and/or CO_2 storage capacity of shales (Ross and Bustin 2009; Wang et al. 2013a, b; Hazra et al. 2015; Hong et al. 2016). It is understandable that with increasing organic richness, the proportion of micropores within shales increase, and thereby improves the gas storage properties of the horizon. On the other hand, several have also reported the lack of role of organic richness on gas storage properties of shales. For example, Gasparik et al. (2012) in one study on shales from Netherlands with TOC varying between 0.8 and 10.5 wt.%, presented lack of role of TOC on CH_4 sorption capacity. Similarly, Varma et al. (2014) also noted similar gas adsorption capacities in Permian shales from India with varying organic matter quantity.

One of the other critical factors, which less clearly understood but strongly influence gas storage capacity of shales is their organic porosity. The presence of porosity hosted by organic matter in shales was first documented by Mann et al. (1994) who used cryo-scanning electron microscopy on Posidonia shales from Germany. However, the significance of organic matter hosted porosity in shale reservoirs for holding hydrocarbons gained wide recognition after Loucks et al. (2009) documented presence of pores in shale organic matter under SEM. These samples, prior to imaging were milled using argon ion to prepare highly smoothened shales surfaces that allowed imaging under SEM. Ever since then, examination of argon ion milled samples under SEM to elaborate shale porosity, has been used by several.

Amongst the different factors that affect gas storage potential of shales and evolution of organic matter hosted porosity, thermal maturity levels of the sedimentary organic matter in shales have been most commonly accepted to be the most important (Modica and Lapierre 2012; Valenza et al. 2013). Over time, using evolved techniques viz. gas adsorption, microscopy, artificial maturation of natural samples, researchers have documented and reasoned that pores within the organic matter are formed due to thermally induced cracking of kerogen and concomitant expulsion of hydrocarbons from the shale kerogen, leading to formation of porous structures (Wang and Reed 2009; Loucks et al. 2012; Chen and Xiao 2014). In addition to thermal maturity, Loucks et al. (2012) also pointed towards the influence of relative quantities of reactive kerogen (or hydrogen rich kerogen) and inert kerogen (or hydrogen poor kerogen), with the former showing formation of pores with increasing thermal maturation and generation-expulsion of hydrocarbons, while the later showing minimal formation of porosity due to little or no hydrocarbon generation-expulsion. Similarly, Liu et al. (2017) in their study on New Albany shales using optical-electron microscopy documented progressive increase in organic pores within solid bitumen with increasing thermal maturity, and cellular macro pores (not dependent on thermal maturity level) within shales across all thermal maturity levels, thereby implying the significance of both organic matter type and thermal maturity. Similarly, the impact of organic matter type and thermal maturity on organic matter hosted shale porosity has also been documented by several others (Jennings and Antia 2013; Fishman et al. 2013; Curtis et al. 2012a; Hazra et al. 2016).

While the influence of thermal maturity on evolution of organic pore in shales has been found to be profound, its impact has been doubted as thermal maturation and burial of OM can simultaneously form and destroy organic matter pores in shales (Milliken et al. 2013). In their work, Milliken et al. (2013) noted organic richness instead of thermal maturity stages to control the development of organic matter hosted porosity. While considering evolution of organic porosity in shale reservoirs in comparison to CBM reservoirs, certain complexities can be observed. Researchers while working on shale petroleum systems have reported presence of porosity in the solid bitumen fraction, and not within the kerogen co-existing with it (Pommer and Milliken 2015; Löhr et al. 2015; Schieber et al. 2016; Liu et al. 2017; İnan et al. 2018), as structured organic matter or kerogen rearranges itself through the mechanism of aromatization taking place as the generated oil is expelled/diffused from its structure (İnan et al. 2018). However, this observation that organic porosity forms within the labile solid bitumen to a large extent is counter-intuitive to the findings and reporting on high rank humic coals' gas adsorption capacity, where the impact of maceral type and thermal maturity on microporosity and gas adsorption capacity has been highlighted (previously discussed in Sect. 6.1) (Clarkson and Bustin 1996; Crosdale et al. 1998; Laxminarayana and Crosdale 1999; Flores 2014). Humic coals are primarily composed of type III and IV kerogen and are generally deficient in H-rich type I and II kerogen, owing to which formation of solid bitumen within them is negligible. Going by the inferences of porosity only with solid bitumen, then it is impossible to explain the gas storage capacity and CBM potentiality of humic coals. Likewise, the significance of structured organic matter type (maceral) in influencing the gas adsorption capacity of type III–IV kerogen bearing Permian shales and coals from India, devoid of any identifiable solid bitumen, has been highlighted by Hazra et al. (2015, 2016) and Varma et al. (2015), respectively. This apparent deviation in observations related to the porosity and gas storage capacity of structured macerals in shales and coals, may be related to the limitation of SEM in mapping smaller pores in kerogen. One of the major limitations of SEM is that it is unable to map or image pores smaller than 5 nm in diameter. In their optical-electron microscopy work on Devonian-Mississippian shales, Liu et al. (2017) only considered pores > 20 nm diameter. Klaver et al. (2015) observed that a significant fraction of pores < 5 nm in Haynesville shales were undercounted when examined under SEM. As the literature indicates, it is very critical to understand that the gases in shale and coal reservoirs are primarily adsorbed in micropores (<2 nm). Hence while dealing with porosity characterization using only microscopy, the microporosity within the organic matter is unstudied. İnan et al. (2018) consequently used the term SEM-visible porosity, and opined these to have limited influence on gas storage properties of unconventional reservoirs. Similar observations and deductions were also made by Shen et al. (2018) in their study on high and post mature shales. It is thus imperative that while detailing organic porosity of coal and shale reservoirs, a combination of techniques should be used. However, clearer understanding of the mechanisms of development of organic porosity in unconventional reservoirs, still presents challenges and different opinions (Katz and Arango 2018).

Similar to thermal maturity and organic richness, the kerogen type present within shales has also been identified to play a critical role in affecting the storage properties of shales. Zhang et al. (2012) in their pioneering work examined CH_4 capacity of bulk shales and isolated kerogens, and at all experimental temperature conditions viz. 35, 50, 60 °C, type III kerogen showed highest sorption capacity for CH_4, followed by type II and type I kerogen. This was attributed to the greater proportions of aromatic structures in type III kerogen, relative to aliphatic structures in type I and II kerogens. Similar observations were reported by Hazra et al. (2015), where they observed higher CH_4 adsorption capacity in vitrinite rich shales. Chalmers and Bustin (2008) in their study on shales from the Lower Cretaceous Buckinghorse Formation, noted that on per unit TOC basis, type II/III admixed kerogen showed highest CH4 sorption capacity, followed by type III, II and I kerogens.

Similar to the enhanced CBM recovery, CO_2 enhanced shale gas recovery (CO_2-ESGR) has received considerable research interests in the past decade. For CO_2-ESGR projects, the competitive adsorption and preferential adsorption of CO_2 over CH_4, enabling additional production of CH_4 and simultaneous sequestration of CO_2 is used. It is estimated that CO_2-ESGR can potentially increase CH_4 production by as much as 26%, and can sequestrate > 60% of injected CO_2 in shale following primary recovery (Godec et al. 2013; Kim et al. 2017; Liu et al. 2018; Wan and Mu 2018; Li et al. 2019). The first examination of CO_2 versus CH_4 adsorption on several Devonian black shales from Kentucky, USA was carried out by Nuttal et al. (2003, 2005), and significantly they observed the absolute mass adsorption of CO_2 to be more than 5 times than CH_4 at similar pressures. Subsequently, several other researchers examined the CO_2 versus CH_4 competitive adsorption in controlled experimental settings, and noted selectively higher adsorption of CO_2 relative to CH_4 in shales (Weniger et al. 2010; Chareonsuppanimit et al. 2012; Heller and Zoback 2014; Pozo et al. 2017). For carbonaceous shales from Paraná Basin, Brazil, Weniger et al. (2010) noted the CO_2:CH_4 adsorption ratio to vary between 1.9 and 6.9, indicating the stronger organic matter affinity in both coals and shales towards CO_2. Kang et al. (2011) observed 5–10 times higher CO_2 adsorption capacity of Barnett shales (USA) than CH_4. Similarly, Zhou et al. (2019a, b) noted upto 8 times higher CO_2 sorption capacity than CH_4 sorption capacity of dried shales. Qi et al. (2019) in their study based on experiments and modelling examined the impact of pore sizes on the CO_2 versus CH_4 competitive adsorption and quiet significantly noted that CO_2 exhibited preferential adsorption in pores with sizes > 0.42 nm. over CH_4, whereas the reverse was observed for pore sizes < 0.42 nm. In general, the competitive adsorption of CO_2 versus CH_4 in shales, is influenced by the interaction energies, molecular sizes and accessibility of CO_2 and CH_4 to the pore structural network in shales (steric forces). It is fairly well established that the dynamic diameter of CO_2 is lower (0.33 nm) in comparison to that of CH_4 (0.38 nm) (Duan et al. 2016), and this essentially allows CO_2 to access greater areas of shale pore systems (Kang et al. 2011).

6.3 SC-CO_2-Based Enhanced Recovery in Unconventional Reservoirs

Injection of CO_2 into deep coal seams or shale reservoirs leads to its phase transition from gaseous CO_2 to supercritical CO_2 (SC-CO_2) when the pressure and temperature conditions reaches the critical point at 7.38 MPa and 31.1 °C, respectively (He et al. 2014; Zhou et al. 2018; Hazra et al. 2022). In recent years numerous studies have been conducted to understand the interaction mechanisms that affects the reservoir properties. In general, SC-CO_2 influences coal seams or shale formations by extracting organic matter, causing swelling through adsorption, and inducing mineral dissolution (Vishal 2017; Pan et al. 2018; Xie et al. 2021; Liu and Liu 2022). These interactions ultimately result in the development of broader and extended fractures thereby enhancing hydrocarbon recovery (Hazra et al. 2022). Through the preferential adsorption of CO_2 and the desorption of CH_4, the advancement of unconventional-play development utilizing SC-CO_2 also offers an opportunity for additional CO_2 sequestration (Middleton et al. 2015; Zhang et al. 2017; Lyu et al. 2021). Furthermore, SC-CO_2 has been observed to be better suited when used during the drilling phase of the development of unconventional reservoirs due to its capability of achieving higher penetration rates in comparison to conventionally employed fluids that are water-based (Kolle 2000; Haizu et al. 2011).

6.3.1 *SC-CO_2-Based Drilling and Fracturing*

SC-CO_2 has emerged as a promising fluid for unconventional reservoir drilling and fracturing, offering unique advantages compared to traditional water-based fluids. Unconventional reservoirs present challenges during drilling, particularly concerning overbalance conditions and formation damage when using conventional water-based fluids. Several studies have highlighted the potential risks associated with these reservoirs, necessitating alternative solutions (Middleton et al. 2015; Burrows et al. 2020). One such solution involves utilizing low-density fluids to lower hydrostatic pressure and mitigate overbalance conditions (Herzhaft et al. 2000; Shahbazi et al. 2007; Maranuk et al. 2014). However, these fluids (e.g., nitrogen foam, light compressed air, or mineral oil) often lack the necessary torque to optimize bottom hole motor efficiency (Qutob and Byrne 2015; Hazra et al. 2022). In contrast, SC-CO_2's liquid-like density provides the required torque for efficient motor operation (Tian et al. 2016). Additionally, SC-CO_2 low viscosity reduces frictional head loss, contributing to high penetration rates and low threshold pressures (Wang et al. 2012). The process of SC-CO_2-based drilling typically involves storing liquid CO_2 in high-pressure tanks and pumping it to the bottom hole using specialized equipment (Shahbazi et al. 2007). As CO_2 is injected into the formation, it undergoes phase transition when the temperature and pressure reach the critical threshold. The depth at which this phase change will occur depends on the unique geothermal gradient of each formation, adding a layer of complexity and adaptability to SC-CO_2 drilling strategies (Middleton et al. 2015). SC-CO_2 fluid offers a significant benefit during drilling as it generates a potent jet when directed through the drill bit, as a result it achieves greater penetration depth and a higher threshold erosion pressure in comparison to water (Kolle 2000). SC-CO_2 dissolve heavy oil components and organic matter near the wellbore, reducing pore throat plugging and enhancing gas and oil flow (Haizu et al. 2011). This dissolution capability is influenced by factors such as temperature, pressure, boiling point, molecular weight, and polarity of the substances involved (Hazra et al. 2022). Recent explorations in SC-CO_2 based drilling technology have been directed towards enhancing our understanding through the lens of mathematical modelling and the principles of computational fluid dynamics (CFD). CFD simulation models help understand how changing the pressure and temperatures at the start affects the speed and power of a supercritical CO_2 jet used in drilling (Long et al. 2016). Escalation in inlet pressure directly correlates with an upsurge in mass flow rate and the jet's impact force. Conversely, the alterations in temperature presented a more complex relationship, indicating the need for a nuanced approach to temperature control. Additionally, the manner in which the SC-CO_2 jet emanating from the drill bit is impacting the formation is correlated to formation erosion efficiency (Huang et al. 2018; Hazra et al. 2022). A thorough comparison of erosion characteristics, including erosion area, depth, intensity, and rate, indicates that SOPSJs demonstrate superior erosion performance when compared to continuous jets (Huang et al. 2018).

SC-CO_2-based drilling also faces challenges such as elastomer compatibility, solids transportation, and proppant carrying capacity (Hazra et al. 2022). CO_2 in supercritical state, acting as a solvent, can induce significant swelling in polymers, necessitating the design of elastomers (a type of polymer used in the transport mechanism of CO_2) that are congruent with supercritical CO_2 conditions (Gupta et al. 2005; Ansaloni et al. 2020). Challenges related to solids transportation during SC-CO_2 drilling, stem from its low viscosity, which hampers the effective removal of drilling debris, known as cuttings (Hazra et al. 2022). To mitigate this problem, studies have shown that additives like fluoroether disulfate telechelic ionomer can be used to enhance the cuttings carrying capacity of SC-CO_2 (Lyu et al. 2021). Also, adjusting the pump displacement to increase flow rate enhances cuttings carrying efficiency by facilitating their movement within the fluid. Managing the incline of the well is also crucial, as it directly impacts the transport effectiveness; tweaking the well's angle optimizes cuttings' flow. Additionally, controlling temperature and pressure within the wellbore plays a pivotal role. Elevating temperature under constant pressure improves carrying ability, while adjusting pressure while maintaining temperature can impact efficiency negatively. These interventions collectively enhance the SC-CO_2 fluid's capacity to transport cuttings effectively, improving overall operational performance (Li et al. 2012). Adjusting the wellhead back pressure to control the density of SC-CO_2 can also aid in improving the fluid's ability to carry cuttings effectively (Shen et al. 2011). These solutions are crucial steps in enabling the wide scale implementation of the SC-CO_2 based drilling technologies (Hazra et al. 2022).

Apart from drilling, SC-CO_2-based fracturing techniques have gained attention for their ability to generate complex fracture networks in unconventional rock formations (Zhang et al. 2017; Hazra et al. 2022). Figure 6.4 shows the variation in the major physical, chemical and mechanical properties of shales caused by SC-CO_2-based fracturing compared to hydraulic fracturing. SC-CO_2 offers distinct advantages over aqueous fluids such as rapid fracture creation and expansion towards boundary, high flowback rates, and minimal formation damage (Wang et al. 2017). The fracturing process with SC-CO_2 compared to water or oil achieve comparatively lower breakdown pressure due to its low viscosity (Ishida et al. 2016; Wang et al. 2018a, b; Ha et al. 2018; Hazra et al. 2022). Table 6.1 lists the various elements that affect SC-CO_2 fracturing based on breakdown pressure. Low-viscosity fluids like liquid CO_2 are key in creating effective fracture networks that enhance stimulated reservoir volume (SRV) (Bennour et al. 2018). Yet, the ability to carry and distribute proppant effectively in SC-CO_2 fracturing is a major challenge that needs addressing to improve this method (Wang and Elsworth 2020; Zhang et al. 2021; Hazra et al. 2022). Proppant plays a vital role in SC-CO_2 fracturing of shale reservoirs by keeping fractures open after the fracturing process (Hazra et al. 2022). This allows for uninterrupted hydrocarbon flow from the reservoir to the wellbore. Without proppant, fractures would close under overburden pressure, making fracturing ineffective. The proppant carrying capacity of SC-CO_2 during fracturing is influenced by several key factors (Chen and Sun 2022). Firstly, the fluid properties of SC-CO_2, such as density and viscosity, are critical and can be adjusted via temperature and pressure control. Proppant characteristics, including size, shape, and density, also play a role, with larger or denser proppant particles settling more quickly. Flow rate affects carrying capacity, as higher velocities can transport proppant deeper into fractures. Fracture geometry, particularly width and complexity, can hinder proppant movement. Perforation hole orientation and density play crucial roles in fracture initiation and propagation during SC-CO_2 fracturing (Chen et al. 2019). Temperature, jet standoff distance, and ambient pressure are key factors that influences SC-CO_2 jet perforation performance (Wang et al. 2018a, b). Additionally, breakdown pressure is observed to decrease with increasing temperature during SC-CO_2 fracturing (Zhou et al. 2019a, b). SC-CO_2-based drilling and fracturing represent innovative approaches with significant

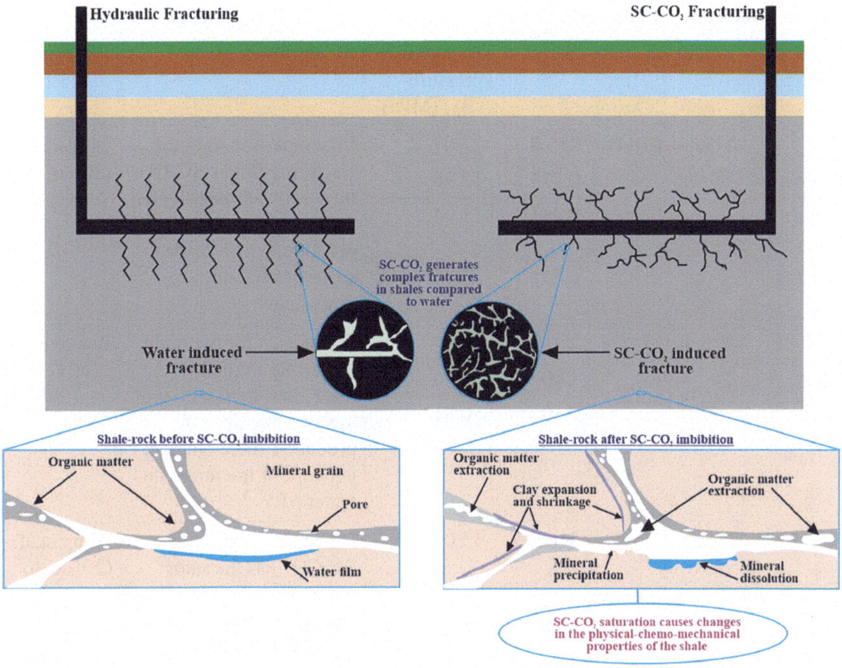

Fig. 6.4 Advantages of $SC\text{-}CO_2$ fracturing versus hydraulic fracturing and its influence on the physical, chemical and mechanical properties of shales. Reprinted (adapted) with permission from Pan et al. (2018); Energy Fuels 2018, 32, 2, 1963–1978. Copyright © 2022 American Chemical Society

potential for enhancing efficiency and reducing environmental impact in unconventional reservoir operations. While challenges such as solids transportation, proppant carrying capacity, and perforation optimization persist, ongoing research and development efforts aim to overcome these obstacles and fully realize the benefits of $SC\text{-}CO_2$ technology in the oil and gas industry.

6.3.2 *Influence of $SC\text{-}CO_2$ on Petrophysical Properties of Coal and Shales*

The interaction of coal and shales with $SC\text{-}CO_2$ can lead to notable changes in their petrophysical properties, particularly porosity and permeability (Hazra et al. 2022). In coal, $SC\text{-}CO_2$ has been observed to increase pore volume and specific surface area, enhancing methane adsorption capacity due to the dissolution of minerals and organic matter (Chen et al. 2023). This process also results in the enlargement of micropores and macropores, which can improve gas flow and recovery. Liu et al. (2023a, b, c) studied the mechanism of adsorption enhancement of coal upon interaction with $SC\text{-}CO_2$-water system. They opted for different reaction times (24, 72, 120, and 240 h) at 9 MPa and 40 °C pressure and temperature, respectively. It was observed

Table 6.1 Aspects that affect SC-CO_2 fracturing based on breakdown pressure

Fracturing fluid	Influencing parameters		Breakdown pressure (B_p) (MPa)	Major observations	References
Oil	Viscosity (mPa s)	316.0	23.07	In comparison to other fracturing fluids, SC-CO_2 fluid's lower viscosity led to a lower breakdown pressure, which made it easier for fractures to propagate	Ishida et al. (2016)
Water		0.774	12.96		
Liquid CO_2		0.097	11.96		
SC-CO_2		0.051	9.10		
Oil	Viscosity (mPa s)	270.00	16.44	When used for hydraulic fracturing in shale reservoirs, fluid with the lowest viscosity can produce a more efficient network of fracture with better SRV	Bennour et al. (2018)
Water		1.00	8.86		
Liquid CO_2		0.10	6.81		
Oil	Dynamic viscosity ($\mu_{\text{fracturing fluid}}$/ Pa s)	316.8×10^{-3}	13.0	Dynamic viscosity of SC-CO_2 has the same effect as static viscosity leading to more effective fracture network formation	Wang et al. (2018a, b)
Water		0.79×10^{-3}	12.8		
SC-CO_2		4.04×10^{-5}	8.44		
Water	Perforation angle (°)	0	14.5	In shales, the perforation angle significantly affects the spread of fractures. It was observed that SC-CO_2 fracturing outperformed water for fracturing at various perforation angles	Chen et al. (2019)
		30	14.27		
		45	14.07		
		60	16.81		
		90	18.01		
SC-CO_2		0	12.89		
		30	13.21		
		45	15.38		
		60	16.03		
		90	16.27		
SC-CO_2	Temperature (°C)	25	20.21	As the temperature rises, SC-CO_2 loses some of its viscosity, resulting in a lower breakdown pressure and easier propagation of fracture	Zhou et al. (2019a, b)
		40	17.82		
		55	16.78		
		70	16.53		

(continued)

Table 6.1 (continued)

Fracturing fluid	Influencing parameters		Breakdown pressure (B_p) (MPa)	Major observations	References
		85	13.36		
		100	13.24		
SC-CO_2	Bedding plane angle (°)	0	40.51	As the bedding plane angle increases, SC-CO_2 fracture becomes simpler. This agrees with the observations of Ha et al. (2018). They discovered that as the orientation angle increased, both SC-CO_2 and freshwater fracture became easier, but the Bp of samples fractured with freshwater was exceeded those fractured using SC-CO_2	Zhang et al. (2019)
		15	36.38		
		30	34.42		
		45	35.89		
		60	32.31		
		75	34.49		
		90	31.09		
SC-CO_2	Injection rate (ml/min)	10	55.42	As the rate of fluid injection increases, fracture formation becomes more challenging	Jia et al. (2021)
		15	52.26		
		20	55.22		

that SSAs for pores with size < 100 nm witnessed an increase of 26–70% from 24 to 240 h after a decrease of 17–38% at the initial 24 h. Chemical reaction induced pore filling resulting in the decrease of SSA and pore volume was also observed. The SSAs (for pores with size > 100 nm) of the coal samples increased by 14–51% from 0 to 240 h after exposure to SC-CO_2-water system. The complicated nature of the impact of SC-CO_2 on coal micropores is further explained by Liu et al. (2019). They observed that the Micropores (<0.46 nm) linked to inter-layer spacing of aromatic layers exhibit increased volume due to SC-CO_2-induced swelling, causing detachment of heterocycles and non-condensed polynuclear aromatics in high-volatile bituminous coal. Conversely, in semi-anthracite and anthracite coal, they observed falling off of low bond energy cross-links and the formation of C_{Ar}–C_{Ar} cross-links through polyaddition reactions result in decreased micropore volumes < 0.46 nm. Micropores (>0.46 nm), identified as intermolecular pores in the coal's macromolecular structure, experience a reduction in volume due to broken bonds caused by SC-CO_2, leading to the formation of new aliphatic or aromatic compounds. With increasing coal rank, SC-CO_2 treatment prioritizes the macromolecular structure, enhancing the volume of micropores in semi-anthracite and anthracite coal. Wang et al. (2023)

studied the influence of SC-CO_2 on the chemical structure and mineral composition of deep seam coals with different tectonic deformation degrees. They observed that with increase in the degree of tectonic deformation, the dissolution impact on aromatic compounds intensifies, while aliphatic compounds exhibit a complementary evolution trend. Supercritical carbon dioxide (SC-CO_2) significantly dissolves carbonate minerals (100%), followed by clay minerals (1.3–30.5%), while oxide minerals remain stable. As tectonic deformation increases, mineral solubility generally rises. SC-CO_2 profoundly affects both organic functional groups and minerals in tectonically deformed coal, its influence growing with higher deformation degrees. Considering the pivotal role of mineral and organic composition in coal porosity and adsorption, injecting CO_2 into deep coal seams requires assessing changes in porosity, permeability, and adsorption due to altered mineral and organic composition.

SC-CO_2 exerts a significant influence on the petrophysical characteristics of shales, particularly in terms of porosity, and permeability (Hazra et al. 2022). Shale porosity encompasses inorganic (resulting from minerals), organic (due to organic matter), and fracture porosity, with pores spanning from micropores to macropores, while permeability governs gas and oil flow in shale formations (Curtis et al. 2012c; Hazra et al. 2018a, b; Arif et al. 2021). SC-CO_2 has the capacity to dissolve organic and mineral components present in the shale, thus creating new fluid flow pathways and enhancing the existing pore network (Jiang et al. 2016; Vishal et al. 2021b). Moreover, SC-CO_2's low viscosity and high diffusivity enable it to penetrate the rock matrix effectively, further elevating permeability (Lyu et al. 2020).

Shale, composed primarily of minerals such as clay, quartz, and calcite, along with small amounts of feldspars, carbonates, and pyrite, undergoes notable changes in composition and morphology upon SC-CO_2 exposure (Ross and Bustin 2009; Ougier-Simonin et al. 2016). Studies comparing the CO_2 and SC-CO_2 saturation of shales highlighted the differential effects on pore characteristics, with SC-CO_2 promoting mineral dissolution while gaseous CO_2 treatment primarily influences swelling and adsorption (Rogelj et al. 2018; Vishal et al. 2021b). SC-CO_2 interaction also causes the release of "crystal water" from clay minerals and dissolution of organic materials, leading to alterations in pore structures and fractures, ultimately enhancing pore-fracture connectivity (Jiang et al. 2016). Analyses of elemental mobility post-SC-CO_2 treatment revealed primary dissolution in carbonates, mobilizing Mg and Ca elements, while clay minerals experience secondary mobilization, mobilizing Al, K, and Na elements (Luo et al. 2019). Similarly, Ozotta et al. (2021) observe changes in mineral content, with increased quartz in Upper Bakken shales and decreased content in Lower Bakken shales, emphasizing the impact of exposure duration. The increase in quartz content was attributed to the relative decrease of clay minerals with increasing SC-CO_2 saturation time. SC-CO_2 exposure also effects pore structure dynamics of shale, such as changes in fractal dimensions, specific surface areas, and overall pore volumes (Hazra et al. 2022). The fractal dimension and specific surface area of the shales were observed to initially increase until 8 and 16 days and then decreased (Ozotta et al. 2021). Wang et al. (2019) observed that SC-CO_2 exposure in Middle Bakken shales primarily impacted mesopores with minimal effects on micropores and macropores. Conversely, Hui et al. (2019) noted variable outcomes in Sichuan and Ordos Basin shales, with specific surface area increasing slightly in oil-window maturity samples but showing significant micropore surface area growth in CO_2-treated samples. Exposure to SC-CO_2 alone had minimal impact on shale mineralogy, unlike SC-CO_2 combined with brine, which led to increased porosity due to pH changes and mineral deposition (Huang et al. 2019).

Another significant petrophysical characteristic that is influenced by SC-CO_2 and has an impact on the reservoir performance, particularly for sequestration, is the surface wettability (Hazra et al. 2022). Wettability plays a crucial role in CO_2 sequestration, affecting how CO_2 interacts with reservoir rock and brine in coal and shale formations. It influences capillary trapping, where a more CO_2-wet reservoir allows efficient structural trapping of CO_2 in pore spaces, enhancing storage capacity and cap rock integrity for long-term containment (Zhu et al. 2011; Ibrahim and Elkatatny 2023). Managing wettability is key to optimizing CO_2 sequestration in unconventional reservoirs, ensuring efficient storage and recovery processes (Hazra et al. 2022). Zhu et al. (2019) explored the wettability changes of coal surfaces due to interactions with gas, liquid, and SC-CO_2 at varying pressures and temperatures. Gaseous CO_2 caused minimal wettability alterations, shifting from water-wet to intermediate-wet. In contrast, liquid and SC-CO_2 reactions resulted in strongly CO_2-wet surfaces, indicated by significantly increased contact angles. Both static and dynamic contact angles rose notably with pressure, while temperature had a reverse effect, especially under supercritical conditions. These trends correlate with CO_2 density variations due to pressure and temperature fluctuations. Cui et al. (2024) investigated the wettability of the coal-water-CO_2 system under different pressures (3–8 MPa) and temperatures (40–70 °C) across four coal ranks using the sessile drop method. They observed a consistent increase in contact angle (CA) with rising pressure up to 5 MPa, followed by a rapid increase at 5–8 MPa, while CA decreases with temperature and coal rank. Xu et al. (2023) studied the impact of CO_2–H_2O-coal interactions on coal wettability using LF-NMR combined with experimental interactions. Their results revealed that water's dynamic wetting in coal primarily hinges on pore structure. Geochemical reactions alter the coal's pore structure, enhancing surface hydrophobicity, thereby influencing water's wetting dynamics. While wetting in large pores remains largely unaffected, wetting in small pores significantly shifts due to increased hydrophobicity. Xu et al. (2024) investigated coal pore structures and water dynamics during wetting with CO_2–H_2O interactions. Their results revealed significant changes in coal pore structure over time. Surface roughness decreased while hydrophobicity increased after short-term and long-term CO_2–H_2O treatment. Capillary forces transformed free water into capillary and adsorbed water, following a linear driving force mass transfer rate law. Applying modified Wenzel model, they observed that wettability on hydrophilic surfaces post-SC-CO_2 treatment, alter coal's contact points and apparent contact area for improved water retention. Ding et al. (2023) investigated coal-water contact angle under different gas environments and analysed coal-water adsorption mechanisms using molecular dynamics simulations. They observed a significant increase in contact angle in CO_2, followed by N_2 and He environments.

Qin et al. (2017) studied the impact of SC-CO_2 on shale wettability, revealing a decrease in water wettability post-treatment. This change facilitated enhanced CO_2 diffusion into the shale matrix, thereby improving its CO_2 adsorption and CH_4

desorption capacities. Similarly, Kaveh et al. (2016) explored the CO_2 wettability of silty shale caprocks and discovered a robust water-wet system suitable for CO_2 storage. Fatah et al. (2021b) examined the wettability of shales with differing mineralogical compositions post SC-CO_2 treatment. They observed a shift from water-wet to CO_2-wet behavior in clay-rich shales, contrasting with quartz-rich shales that maintained water-wet characteristics. In a broader analysis comparing Wolfcamp, Mancos, and Eagle Ford shales, Fatah et al. (2021a) noted that pressure and treatment duration influenced CO_2 wettability differently based on mineralogy. Shales rich in clay showed increased CO_2 wettability due to mineral dissolution, while quartz-rich shales remained water-wet. Guiltinan et al. (2017) explored SC-CO_2 effects on shale wettability across varying thermal maturities and total organic carbon (TOC) concentrations. They found that despite changes in organic content and maturity, shale wettability remained predominantly water-wet, highlighting the dominance of shale mineralogy in wetting behavior. Organic-rich caprock reservoirs could be advantageous for CO_2 sequestration, provided the organic content remains below the percolation threshold to avoid altering wetting behavior (Hazra et al. 2022). Liu et al. (2023a, b, c) studied the stability of SNF-formed nanoparticle layers on Longmaxi and Yanchang shale under SC-CO_2, investigating wettability reversal at varying nanofluid aging times and concentrations. Results revealed stable nanoparticle structures in SC-CO_2, rendering shale more hydrophilic, especially in quartz-rich Longmaxi shale. Longmaxi samples treated with SC-CO_2 and SNF showed more pore transformations compared to solely SC-CO_2-treated samples, while Yanchang samples exhibit minimal pore structure changes after SNF treatment. The capillary force recovery difference suggests that silica nanofluid may benefit Longmaxi formations more in CO_2-geo-storage applications. Yang et al. (2024) water wettability alterations in shale under varying SC-CO_2-water exposure pressures and temperatures. They observed reduction in water wettability post-SC-CO_2-water exposure, with the extent of reduction influenced by exposure pressure and temperature. The reduction was attributed to changes in shale mineral composition, element mobilization, and pore structure, notably involving increased hydrophobic quartz content, decreased hydrophilic clay and carbonate mineral contents, and a shift in pore sizes from micro and mesopores to larger ones. These alterations collectively contributed to the weakening of water wettability in shale, potentially impacting CO_2 sweep efficiency and capillary trapping capacity in shale.

6.3.3 *Influence of SC-CO_2 on Geomechanical Properties of Coal and Shales*

SC-CO_2 exerts significant influence on the geomechanics of unconventional reservoirs, impacting both hydrocarbon extraction and carbon sequestration endeavours (Hazra et al. 2022). When injected, SC-CO_2 can induce fracturing, enhance permeability, and alter stress distributions within the rock formation, affecting wellbore integrity and seismic activity (Hazra et al. 2022). SC-CO_2 acts as a carrier fluid, aiding hydrocarbon movement through rock pores by dissolving or displacing them from tight matrices found in formations like shale plays or tight gas reservoirs. This enhances recovery efficiency by mobilizing trapped hydrocarbons, facilitating their flow to production wells. However, these changes in the rock matrix also impact geomechanical stability. SC-CO_2 injection also holds promise for carbon sequestration, but its long-term containment within storage reservoirs requires careful geomechanical assessment (Hazra et al. 2023). Several studies have shown the impact of SC-CO_2 interaction on the strength and elastic properties of coal and shales (Perera et al. 2013; Liu et al. 2023a, b, c; Wang et al. 2024). The behaviour of CO_2 injected in coal seams for enhanced coal bed methane recovery (ECBM) and storage is complicated because of its chemical reactivity with the coal matrix and the resulting changes to the chemical and physical structure of the coal mass (Perera 2018). A study by Perera et al. (2013) comparing the impact of gaseous CO_2, SC-CO_2, and N_2 saturation on the mechanical properties of bituminous coal revealed that SC-CO_2 has greater influence on the strength reduction of coal compared to gaseous CO_2 and N_2. Interaction of coal with gaseous CO_2 reduced the UCS and Young's modulus of coals by 53% and 36%, respectively, whereas interaction with SC-CO_2 resulted in decrease of 93% of UCS and 72% of Young's modulus. Thus, the interaction of SC-CO_2 with coal can substantially diminish the mechanical integrity of the designated coal reservoir. In another study, Perera et al. (2013) observed that gaseous CO_2 causes up to 53% UCS reduction and 36% Young's modulus reduction, while SC-CO_2 leads to 40% greater UCS reduction and 100% greater Young's modulus reduction compared to gaseous CO_2. The significant impact of supercritical CO_2 is attributed to its greater adsorptive potential, coal swelling, and dissolution, leading to coal plasticization. Meng and Qiu (2018) in their study related this macroscopic phenomenon of SC-CO_2-induced strength reduction with the changes in the microscopic properties of coal. They conducted several microscopic quantitative experiments such as SEM, nuclear magnetic resonance (NMR), and mercury injection porosimetry (MIP). It was observed that SC-CO_2 treatment resulted in increase in both the diffusion space consisting of micropores, and the percolation space involving the macropores and fractures. The enlargement of pores spaces was attributed to the ability of SC-CO_2 in dissolving low molecular weight organics such as n-alkenes and polycyclic aromatic compounds. Additionally, when mixed with the formation water SC-CO_2 can form carbonic acid that erodes the coal skeleton. Li et al. (2019) studied the influence of SC-CO_2 interaction time (12 and 48 h) on the changes in the mechanical properties of coal under triaxial conditions. At higher confining pressures the coal samples failed with higher compressive strength, however the strain was observed to show a transition from brittle to ductile with increasing SC-CO_2 treatment time. Su et al. (2020) studied the induced changes in the mechanical properties and pore structures of coal after cyclic and continuous saturation with SC-CO_2. The cyclic saturation process of the coal samples with SC-CO_2 for a 1-day period involved saturation at 8 MPa for 12 h and then at 4 MPa (after slow venting) for another 12 h. On the other hand, sustained SC-CO_2 saturation of the coal samples for a 1-day period was conducted at 8 MPa for 24 h. Both cyclic and sustained saturation processes on the coal samples were conducted for 1-, 5-, 9-, and 13-day periods. The treated samples were studied for their UCS, and the pore-structure changes were recorded with the use of NMR and AE. It was observed that the cyclically saturated coal samples showed more strength reduction compared to the continuously saturated samples. It was attributed to the mechanical fatigue due to cyclic saturation. Additionally, cyclic saturation also had greater influence on the pore structure compared to the sustained saturation conducted on coal. Essentially, the increase in the number of saturation cycles leads to need of low tensile stress for the new defects to form in coal. Investigation of cyclic SC-CO_2 interaction with coal is a critical aspect in the understanding of SC-CO_2-induced changes in coal properties because in field conditions sometimes the well has to stop its operation due to unavailability of CO_2 and remains so until it becomes available again, which in essence is a cyclic injection process (Pan et al. 2018; Meng and Qiu 2018). Liu et al. (2023a, b, c) studied the impact of SC-CO_2 saturation pressure on the mechanical properties of coal. It was observed that Mechanical strength decreases notably (up to 35.6%) with increasing saturation pressure due to mineral dissolution. Similarly, Wang et al. (2024) through simulation studies observed that SC-CO_2 increases coal deterioration, promotes plastic failure, and accelerates combined body failure. Yang et al. (2023a, b) investigated SC-CO_2 and SC-CO_2-water interactions on coal microstructures and mechanical strength. SC-CO_2-water reactions significantly impact coal properties, with water playing a major role in altering pore structure, functional groups, and mineral compositions. SC-CO_2 saturation alone induces a modest decline in aliphatic groups and a slight reduction in uniaxial compressive strength (UCS). However, SC-CO_2-water saturation leads to substantial extraction of aliphatic groups and a significant decrease in UCS due to accelerated hydrocarbon extraction and mineral dissolution.

The impact of SC-CO_2 on the mechanical properties of shales has garnered significant attention in recent years (Hazra et al. 2022). Several studies have delved into the intricate dynamics of CO_2 interactions with shales, unveiling multifaceted outcomes (Yin et al. 2017; Zou et al. 2018; Lyu et al. 2020; Shi et al. 2020). Significant alterations have been noted in the strength properties of shale following saturation with SC-CO_2, with particular emphasis on the phase-state and gas pressure of CO_2 as crucial factors determining the magnitude of mechanical changes in shales. The duration of saturation also plays a pivotal role (Lyu et al. 2018; Choi et al. 2021). Prolonged exposure to SC-CO_2 intensifies the weakening effect, thereby accentuating the mechanical degradation of shales over time (Fig. 6.5). This weakening effect is attributed to SC-CO_2 induced pore structure alterations, development of new pores, alterations in carbon content, and mineral dissolution effects causing changes resulting in overall reductions in elastic modulus, tensile strength, and triaxial compressive strength of shales (Ao et al. 2017; Feng et al. 2019). Table 6.2 summarises the changes in the strength and elastic characteristics of shales with respect to the duration of SC-CO_2 treatment. Shales exhibit spatial variability in mineralogical composition, leading to differences in mechanical properties influenced by mineral phases (King 2010; Veytskin et al. 2017; Song et al. 2021). Shi et al. (2020) used nanoindentation techniques to compare the mechanical strengths of quartz-rich and clay-rich regions in shales before and after SC-CO_2 saturation.

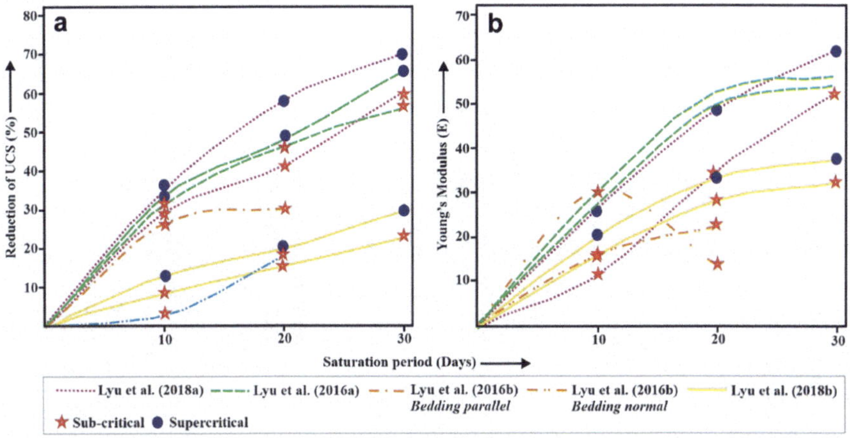

Fig. 6.5 After saturation in subcritical and supercritical CO_2, there is a reduction in **a** UCS-uniaxial compressive strength and **b** E-elastic modulus. Reprinted (adapted) with permission from Hazra et al. (2022); Energy Fuels 2022, 36, 17, 9882–9903. Copyright © 2022 American Chemical Society

They observed significant strength reductions in clay-rich areas due to mineral dissolution and CO_2 adsorption, contrasting with minimal changes in quartz-rich regions impacted less by CO_2 exposure. Yang et al. (2023a, b) investigated the impact of SC-CO_2-water soak at varying pressures (P = 0, 10, 15, 20 MPa) and temperatures (T = 308, 323, 338, 353 K) on mechanical properties shales and observed that SC-CO_2-water soak led to reduced clay and carbonate mineral content, increased macropore volume and porosity, and decreased triaxial compressive strength and elastic modulus of shale. The mechanical weakening was more pronounced with higher soak pressures and lower temperatures. Khosravi et al. (2023) studied the impact of SC-CO_2 exposure on Middle Bakken shale core plugs over varying durations (0, 8, 16, and 30 days) and confining pressures (10, 20, 30, and 40 MPa). Mechanical properties (peak strength, Young's modulus, Poisson's ratio) and acoustic velocities (VP, VS) were measured alongside mineralogical and microstructural changes using XRD and electron microscopy. Fluctuations were observed throughout the exposure period, with an eventual increase in peak strength but decreases in Young's moduli and Poisson's ratios. The initial decline in strength during the first 8 days was primarily caused by clay swelling, mineral dissolution, and crack propagation triggered by carbon dioxide. Interestingly, even minor chemical degradation had a greater impact on peak strength than the effects of confining pressure during this initial period. Subsequently, the confining pressure was elevated to 30 and 40 MPa at 16- and 30-days post-exposure, respectively, resulting in a noticeable increase in peak strength. This adjustment in confining pressure aimed to mimic stress fluctuations expected during carbon capture and storage (CCS) and enhanced oil recovery (EOR) operations due to the presence of supercritical CO_2 in the reservoir.

Table 6.2 Summary of the impact of the duration of the SC-CO_2 treatment on the mechanical characteristics of shales

Sampling area	Treatment time	TOC (%)	Original mineralogical composition before exposure to SC-CO_2	Maximum reduction in strength properties (% reduced) with respect to SC-CO_2 treatment time (h)		Notable changes in mineralogy after SC-CO_2 treatment	Literature
Black shales, Southeast Chongqing, China	10, 20, and 30 D	2–4	Qtz > F > Muscovite > C > Pyrite > Smectite > Illite > Annite > Kaolinite > Braunite	E	56.32% after 30 days	The percentages of C and Fe increased, while K and Al decreased, according to the EDS analysis	Lyu et al. (2016)
				UCS	66.05% after 30 days		
Longmaxi formation, Sichuan Basin, China	24, 48, 72, 96, and 120 h	7.88	Qtz > F > P > C > D > Pyrite > Clay	E	29.56% after 120 h	After 96 h, the quartz content increased. All other shale minerals saw a decline in mineral content as treatment time increased	Ao et al. (2017)
				TS	27.7% after 96 h		
				TCS	15.3% after 120 h		
Silurian Longmaxi formation, Schuan Basin, China	10 D	3.7	Qtz > D > C > F > Clay	E	31.49% after 10 D	There was no discernible difference in the mineral content	Yin et al. (2017)
				UCS	28.67% after 10 D		
Changing region, Sichuan basin, China	10, 30, and 60 D	3.8	Qtz > C > G > I > D	E	46% after 60 days	Mineral content changes were not considered in the study	Feng et al. (2019)
				TS	22% after 60 days		

(continued)

Table 6.2 (continued)

Sampling area	Treatment time	TOC (%)	Original mineralogical composition before exposure to $SC\text{-}CO_2$	Maximum reduction in strength properties (% reduced) with respect to $SC\text{-}CO_2$ treatment time (h)		Notable changes in mineralogy after $SC\text{-}CO_2$ treatment	Literature
Sichuan Basin, China	10, 20, and 30 D	3.8	Qtz > F > Cristobalite > Kaolinite > Illite	E	37.79% after 30 days	Following CO_2 adsorption, the carbon content increased	Lyu et al. (2018)
				UCS	29.95% after 30 days		
Silurian Longmaxi formation, Chongqing, Sichuan Basin, China	10 D	–	Qtz > D > C > Clay > F > Pyrite > Hematite	E	5.7% after 10 D	All shale minerals, with the exception of quartz, saw a decrease in mineral content following $SC\text{-}CO_2$ treatment. The decrease in tensile strength resulted from calcite dissolution	Lyu et al. (2021)
				TS	11.9% after 10 D		
Longmaxi shale, Sichuan basin, China	48 h	7.8	Qtz > F > P > C > D > Pyrite > Clay	FT	11.3% after 48 h	Almost no change in quartz content	Shi et al. (2020)
				E	11.0% after 48 h		
				H	29.5% after 48 h		

(continued)

Table 6.2 (continued)

Sampling area	Treatment time	TOC (%)	Original mineralogical composition before exposure to SC-CO_2	Maximum reduction in strength properties (% reduced) with respect to SC-CO_2 treatment time (h)		Notable changes in mineralogy after SC-CO_2 treatment	Literature
Songliao basin, China	3, 7, and 14 D	–	Qtz > F > Cb > Clay > P > Siderite	E	47% after 14 days	Following a 14-day period, there was a 26% increase in quartz content.In a span of 14 days, carbonate minerals exhibited a 12% reduction	Meng et al. (2021)
Longmaxi formation shale, Sichuan Basin, Yibin City, Sichuan Province	15, and 30 D	3.7	Brittle minerals > clay minerals > carbonate minerals	TS	57.5% after 30 days	EDS analysis revealed a decrease in the mineral content of carbonate and silicate	Qin et al. (2021)
Silurian Longmaxi formation, southeast Chongqing, China	2 h @ 30 MPa and 335.15 K	–	Qtz > Clay > C > D	PR	40.3% after 2 h	Variations in mineral content were not considered in the study	Song et al. (2021)
				UCS	22.9% after 2 h		

Reprinted (adapted) with permission from Hazra et al. (2022); Energy Fuels 2022, 36, 17, 9882–9903. Copyright © 2022 American Chemical Society

References

Ahn CH, Chang OC, Dilmore R, Wang JY (2014) A hydraulic fracture network propagation model in shale gas reservoirs: parametric studies to enhance the effectiveness of stimulation. In: SPE/AAPG/SEG unconventional resources technology conference. URTEC

Ansaloni L, Alcock B, Peters TA (2020) Effects of CO_2 on polymeric materials in the CO_2 transport chain: a review. Int J Greenhouse Gas Control 94:102930

Ao X, Lu Y, Tang J, Chen Y, Li H (2017) Investigation on the physics structure and chemical properties of the shale treated by supercritical CO_2. J CO2 Utilizat 20:274–281

Arif M, Mahmoud M, Zhang Y, Iglauer S (2021) X-ray tomography imaging of shale microstructures: a review in the context of multiscale correlative imaging. Int J Coal Geol 233:103641

Aringhieri R (2004) Nanoporosity characteristics of some natural clay minerals and soils. Clays Clay Miner 52:700–704

Arri LE, Yee D, Morgan WD, Jeansonne MW (1992) Modeling coalbed methane production with binary gas sorption. In: SPE proceedings, rocky mountain regional meeting and exhibition, May 18–21, 1991, Casper, WY, pp 459–472

Bakshi T, Prusty BK, Pathak K, Pal SK (2018) Pore characteristics of Damodar valley shale and their effect on gas storage potential. J Pet Sci Eng 162:725–735

Bennour Z, Watanabe S, Chen Y, Ishida T, Akai T (2018) Evaluation of stimulated reservoir volume in laboratory hydraulic fracturing with oil, water and liquid carbon dioxide under microscopy using the fluorescence method. Geomech Geophys Geo-Energy Geo-Resour 4:39–50

Burrows LC, Haeri F, Cvetic P, Sanguinito S, Shi F, Tapriyal D, Goodman A, Enick RM (2020) A literature review of CO_2, natural gas, and water-based fluids for enhanced oil recovery in unconventional reservoirs. Energy Fuels 34(5):5331–5380

Busch A, Gensterblum Y (2011) CBM and CO_2-ECBM related sorption processes in coal: a review. Int J Coal Geol 87(2):49–71

Busch A, Gensterblum Y, Krooss BM, Siemons N (2006) Investigation of high-pressure selective adsorption/desorption behaviour of CO_2 and CH_4 on coals: an experimental study. Int J Coal Geol 66(1–2):53–68

Bustin RM, Clarkson CR (1998) Geological controls on coalbed methane reservoir capacity and gas content. Int J Coal Geol 38(1–2):3–26

Ceglarsk-Stefańska G, Zarębska K (2002) The competitive sorption of CO_2 and CH_4 with regard to the release of methane from coal. Fuel Process Technol 77–78:423–429

Chalmers GRL, Bustin RM (2007) The organic matter distribution and methane capacity of the Lower Cretaceous strata of northeastern British Columbia. Int J Coal Geol 70:223–239

Chalmers GR, Bustin RM (2008) Lower Cretaceous gas shales in northeastern British Columbia, part I: geological controls on methane sorption capacity. Bull Can Pet Geol 56(1):1–21

Chalmers GRL, Bustin RM, Power IM (2012) Characterization of gas shale pore systems by porosimetry, pycnometry, surface area, and field emission scanning electron microscopy/transmission electron microscopy image analyses: examples from the Barnett, Woodford, Haynesville, Marcellus, and Doig Units. AAPG Bull 96(6):1099–1119. https://doi.org/10.1306/10171111052

Chalmers GRL, Bustin RM, Power IM (2009) A pore by any other name would be as small: the importance of meso- and microporosity in shale gas capacity. AAPG Annual Convention and Exhibition, Denver, p 1

Chandra D, Vishal V, Debbarma A, Banerjee S, Pradhan SP, Mishra MK (2020) Role of composition and depth on pore attributes of Barakar Formation Gas Shales of Ib Valley, India, using a combination of low-pressure sorption and image analysis. Energy Fuels 34:8085–8098

Chandra D, Vishal V, Bahadur J, Agrawal AK, Das A, Hazra B, Sen D (2022) Nano-scale physicochemical attributes and their impact on pore heterogeneity in shale. Fuel 314:123070

Chandra D, Bakshi T, Bahadur J, Hazra B, Vishal V, Kumar S, Sen D, Singh TN (2023) Pore morphology in thermally-treated shales and its implication on CO_2 storage applications: a gas sorption, SEM, and small-angle scattering study. Fuel 331:125877

Chareonsuppanimit P, Mohammad SAR, Robert L, Robinson RL, Gasem KAM (2012) High-pressure adsorption of gases on shales: measurements and modeling. Int J Coal Geol 95:34–46. https://doi.org/10.1016/j.coal.2012.02.005
Chen D, Sun Z (2022) Numerical simulation of the proppant settlement in SC-CO_2 sand-carrying fluid in fracturing fractures. Energies 16(1):11
Chen J, Xiao XM (2014) Evolution of nanoporosity in organic-rich shales during thermal maturation. Fuel 129:173–181. https://doi.org/10.1016/j.fuel.2014.03.058
Chen M, Jiang H, Zhang GQ, Jin Y (2010) The experimental investigation of fracture propagation behavior and fracture geometry in hydraulic fracturing through oriented perforations. Petrol Sci Technol 28(13):1297–1306
Chen S, Tao S, Tang D, Xu H, Li S, Zhao J, Jiang Q, Yang H (2017) Pore structure characterization of different rank coals using N_2 and CO_2 adsorption and its effect on CH_4 adsorption capacity: a case in Panguan syncline, Western Guizhou, china. Energy Fuel 31:6034–6044
Chen H, Hu Y, Kang Y, Cai C, Liu J, Liu Y (2019) Fracture initiation and propagation under different perforation orientation angles in supercritical CO_2 fracturing. J Petrol Sci Eng 183:106403
Cheng AL, Huang WL (2004) Selective adsorption of hydrocarbon gases on clays and organic matter. Org Geochem 35:413–423
Choi CS, Kim J, Song JJ (2021) Analysis of shale property changes after geochemical interaction under CO_2 sequestration conditions. Energy 214:118933
Clarkson CR, Bustin RM (1996) Variation in micropore capacity and size distribution with composition in bituminous coal of the Western Canadian Sedimentary Basin: implications for coalbed methane potential. Fuel 75:1483–1498
Clarkson CR, Bustin RM (1999) The effect of pore structure and gas pressure upon the transport properties of coal: a laboratory modeling study. 2. Adsorption rate modeling. Fuel 78:1345–1362
Clayton JL (1998) Geochemistry of coalbed gas—a review. Int J Coal Geol 35:159–173
Crosdale PJ, Beamish BB, Valix M (1998) Coalbed methane sorption related to coal composition. Int J Coal Geol 35:147–158
Crosdale P, Beamish B (1993) Maceral effects on methane sorption by coal. In: Beston JW (ed) New developments in coal geology, a symposium, Brisbane, pp 95–98
Cui X, Bustin RM, Dipple G (2004) Selective transport of CO_2, CH_4, and N_2 in coals: insights from modeling of experimental gas adsorption data. Fuel 83(3):293–303
Cui C, Yao Y, Liu D, Sun X (2024) Decoding wettability in coal-water-CO_2 system for enhanced sequestration security. Gas Sci Eng 12:205285
Curtis JB (2002) Fractured shale-gas systems. AAPG Bull 86(11):1921–1938
Curtis ME, Cardott BJ, Sondergeld CH, Rai CS (2012a) Development of organic porosity in the Woodford Shale with increasing thermal maturity. Int J Coal Geol 103:26–31
Curtis ME, Sondergeld CH, Ambrose RJ, Rai CS (2012b) Microstructural investigation of gas shales in two and three dimensions using nanometer-scale resolution imaging. AAPG Bull 96:665–677
Curtis ME, Cardott BJ, Sondergeld CH, Rai CS (2012c) The development of organic porosity in the Woodford Shale related to thermal maturity. In: SPE annual technical conference and exhibition. SPE, p 160158
Day S, Duffy G, Sakurovs R, Weir S (2008) Effect of coal properties on CO_2 sorption capacity under supercritical conditions. Int J Greenhouse Gas Control 2(3):342–352
Ding N, Si L, Wei J, Jiang W, Zhang J, Liu Y (2023) Study on coal wettability under different gas environments based on the adsorption energy. ACS Omega 8(24):22211–22222
Duan S, Gu M, Du X, Xian X (2016) Adsorption equilibrium of CO_2 and CH_4 and their mixture on Sichuan Basin shale. Energy Fuels 30(3):2248–2256. https://doi.org/10.1021/acs.energyfuels.5b02088
Dutta P, Bhowmik S, Das S (2011) Methane and carbon dioxide sorption on a set of coals from India. Int J Coal Geol 85(3–4):289–299
Ettinger I, Eremin B, Zimakov S, Yavovskaya M (1966) Natural factors influencing coal sorption properties: I. Petrography and the sorption properties of coals. Fuel 45:267–275

Faiz MM, Aziz NI, Hutton AC, Jones BG (1992) Porosity and gas sorption capacity of some eastern Australian coals in relation to coal rank and composition. Proc Symp Coalbed Methane Res Dev Aust 4:9–20

Faiz M, Saghafi A, Sherwood N, Wang I (2007) The influence of petrological properties and burial history on coal seam methane reservoir characterisation, Sydney Basin, Australia. Int J Coal Geol 70(1–3):193–208

Fallahzadeh SH, Shadizadeh SR, Pourafshary P (2010) Dealing with the challenges of hydraulic fracture initiation in deviated-cased perforated boreholes. In: Trinidad and Tobago energy resources conference. OnePetro

Fan E, Tang S, Zhang C, Guo Q, Sun C (2014) Methane sorption capacity of organics and clays in high-over matured shale-gas systems. Energy Explor Exploit 32(6):927–942. https://doi.org/10.1260/0144-5987.32.6.927

Fatah A, Bennour Z, Mahmud HB, Gholami R, Hossain M (2021a) Surface wettability alteration of shales exposed to CO_2: implication for long-term integrity of geological storage sites. Int J Greenhouse Gas Control 110:103426

Fatah A, Mahmud HB, Bennour Z, Hossain M, Gholami R (2021b) Effect of supercritical CO_2 treatment on physical properties and functional groups of shales. Fuel 303:121310

Feng G, Kang Y, Sun ZD, Wang XC, Hu YQ (2019) Effects of supercritical CO_2 adsorption on the mechanical characteristics and failure mechanisms of shale. Energy 173:870–882

Fishman N, Guthrie J, Honarpour M (2013) The stratigraphic distribution of hydrocarbon storage and its effect on producible hydrocarbons in the Eagle Ford Formation, South Texas. In: Unconventional resources technology conference (1540–1545) society of exploration geophysicists, American Association of Petroleum Geologists, Society of Petroleum Engineers

Flores MR (2014) Coal and coalbed gas: outlook. In: Flores RM (ed) Coal and coalbed gas: fueling the future. Elsevier, Amsterdam, pp 587–614. https://doi.org/10.1016/B978-0-12-396972-9.00004-5.

Gan H, Nandi SP, Walker PL (1972) Nature of the porosity in American coals. Fuel 51:272–277

Gasparik M, Ghanizadeh A, Bertier P et al (2012) High-pressure methane sorption isotherms of black shales from the Netherlands. Energy Fuels 26(8):4995–5004. https://doi.org/10.1021/ef300405g

Gentzis T (2000) Subsurface sequestration of carbon dioxide—an overview from an Alberta (Canada) perspective. Int J Coal Geol 43(1–4):287–305

Godec M, Koperna G, Petrusak R, Oudinot A (2013) Potential for enhanced gas recovery and CO_2 storage in the Marcellus shale in the Eastern United States. Int J Coal Geol 118:95–104

Greaves KH, Owen LB, McLennan JD (1993) Multi-component gas adsorption–desorption behavior of coal. In: Proceedings of the 1993 international coalbed methane symposium, University of Alabama, Tuscaloosa, pp 197–205

Gregg SJ, Sing KSW (1982) Adsorption, surface area and porosity. Academic Press, New York

Guiltinan EJ, Cardenas MB, Bennett PC, Zhang T, Espinoza DN (2017) The effect of organic matter and thermal maturity on the wettability of supercritical CO_2 on organic shales. Int J Greenhouse Gas Control 65:15–22

Gupta AP, Gupta A, Langlinais J (2005) Feasibility of supercritical carbon dioxide as a drilling fluid for deep underbalanced drilling operations. In: SPE annual technical conference and exhibition. SPE, p 96992

Ha SJ, Choo J, Yun TS (2018) Liquid CO_2 fracturing: effect of fluid permeation on the breakdown pressure and cracking behavior. Rock Mech Rock Eng 51:3407–3420

Hackley PC, Cardott BJ (2016) Application of organic petrography in North American shale petroleum systems: a review. Int J Coal Geol 163:8–51

Harpalani S, Chen G (1997) Influence of gas production induced volumetric strain on permeability of coal. Geotech Geol Eng 15:303–325

Harpalani S, Schraufnagel RA (1990) Shrinkage of coal matrix with release of gas and its impact on permeability of coal. Fuel 69(5):551–556

Harpalani S, Prusty BK, Dutta P (2006) Methane/CO_2 sorption modeling for coalbed methane production and CO_2 Sequestration. Energy Fuels 20:1591–1599

Hazra B, Varma AK, Bandopadhyay AK, Mendhe VA, Singh BD, Saxena VK, Samad SK, Mishra DK (2015) Petrographic insights of organic matter conversion of Raniganj basin shales, India. Int J Coal Geol 150:193–209

Hazra B, Varma AK, Bandopadhyay AK, Chakravarty S, Buragohain J, Samad SK, Prasad AK (2016) FTIR, XRF, XRD and SEM characteristics of Permian shales, India. J Nat Gas Sci Eng 32:239–255

Hazra B, Wood DA, Kumar S, Saha S, Dutta S, Kumari P, Singh AK (2018a) Fractal disposition, porosity characterization and relationships to thermal maturity for the Lower Permian Raniganj basin shales, India. J Nat Gas Sci Eng 59:452–465

Hazra B, Wood DA, Vishal V, Varma AK, Sakha D, Singh AK (2018b) Porosity controls and fractal disposition of organic-rich Permian shales using low-pressure adsorption techniques. Fuel 220:837–848

Hazra B, Chandra D, Singh AK, Varma AK, Mani D, Singh PK, Boral P, Buragohain J (2019a) Comparative pore structural attributes and fractal dimensions of lower Permian organic-matter-bearing sediments of two Indian basins: inferences from nitrogen gas adsorption. Energy Sour A Recov Utiliz Environ Effects 41(24):2975–2988

Hazra B, Wood DA, Mani D, Singh PK, Singh AK (2019b) Evaluation of shale source rocks and reservoirs, vol 142. Springer, Berlin

Hazra B, Wood DA, Mani D, Singh PK, Singh AK (2019c) Organic and inorganic porosity, and controls of hydrocarbon storage in shales. Evaluation of shale source rocks and reservoirs. Springer, Cham, pp 107–138

Hazra B, Vishal V, Singh DP (2020) Applicability of low-pressure CO_2 and N_2 adsorption in determining pore attributes of organic-rich shales and coals. Energy Fuels 35(1):456–464

Hazra B, Vishal V, Sethi C, Chandra D (2022) Impact of supercritical CO_2 on shale reservoirs and its implication for CO_2 sequestration. Energy Fuels 36(17):9882–9903

Hazra B, Chandra D, Lahiri S, Vishal V, Sethi C, Pandey JK (2023) Pore evolution during combustion of distinct thermally mature shales: insights into potential in situ conversion. Energy Fuels 37(18):13898–13911

He LIU, Feng WANG, Zhang J, Siwei MENG, Yongwei DUAN (2014) Fracturing with carbon dioxide: application status and development trend. Petrol Explor Develop 41(4):513–519

Heller R, Zoback M (2014) Adsorption of methane and carbon dioxide on gas shale and pure mineral samples. J Unconv Oil Gas Resour 8:14–24. https://doi.org/10.1016/j.juogr.2014.06.001

Herzhaft B, Toure A, Bruni F, Saintpere S (2000) Aqueous foams for underbalanced drilling: the question of solids. In: SPE annual technical conference and exhibition. SPE, p 62898

Holmes R, Rupp E, Vishal V, Wilcox J (2017) Selection of shale preparation protocol and outgas procedures for applications in low-pressure analysis. Energy Fuels 31(9):9043–9051

Hong L, Jain J, Romanov V, Lopano C, Disenhof C, Goodman A, Hedges S, Soeder D, Sanguinito S, Dilmore R (2016) An investigation of factors affecting the interaction of CO_2 and CH_4 on shale in Appalachian Basin. J Unconv Oil Gas Resour 14:99–112. https://doi.org/10.1016/j.juogr.2016.02.003

Hou L, Sun B, Geng X, Jiang T, Wang Z (2017) Study of the slippage of particle/supercritical CO_2 two-phase flow. J Supercrit Fluids 120:173–180

Huang M, Kang Y, Wang X, Hu Y, Cai C, Liu Y, Chen H (2018) Experimental investigation on the rock erosion characteristics of a self-excited oscillation pulsed supercritical CO_2 jet. Appl Therm Eng 139:445–455

Huang X, Zhao YP, Wang X, Pan L (2019) Adsorption-induced pore blocking and its mechanisms in nanoporous shale due to interactions with supercritical CO_2. J Petrol Sci Eng 178:74–81

Hui D, Pan Y, Luo P, Zhang Y, Sun L, Lin C (2019) Effect of supercritical CO_2 exposure on the high-pressure CO_2 adsorption performance of shales. Fuel 247:57–66

Ibrahim AF, Elkatatny S (2023) Data-driven models to predict shale wettability for CO_2 sequestration applications. Sci Rep 13(1):10151

İnan S, Al Badairy H, İnan T, Al Zahrani A (2018) Formation and occurrence of organic matter-hosted porosity in shales. Int J Coal Geol 199:39–51

Ishida T, Chen Y, Bennour Z, Yamashita H, Inui S, Nagaya Y, Naoi M, Chen Q, Nakayama Y, Nagano Y (2016) Features of CO_2 fracturing deduced from acoustic emission and microscopy in laboratory experiments. J Geophys Res Solid Earth 121(11):8080–8098

Jarvie DM, Hill RJ, Ruble TE, Pollastro RM (2007) Unconventional shale gas systems: the Mississippian Barnett Shale of north-central Texas as one model for thermogenic shale-gas assessment. AAPG Bull 91(4):475–500

Jarvie D, Hill R, Pollastro R et al (2003) Evaluation of unconventional natural gas prospects: the barnett shale fractured shale gas model. In: Abstracts of poster presented at the 21st international meeting on organic geochemistry, September 8–12, 2003, Krakow

Jennings DS, Antia J (2013) Petrographic characterization of the Eagle Ford Shale, South Texas: mineralogy, common constituents, and distribution of nanometer-scale pore types. In: Camp W, Diaz E, Wawak B (eds) Electron microscopy of shale hydrocarbon reservoirs. AAPG Memoir, pp 101–113

Ji L, Zhang T, Milliken KL, Qu J, Zhang X (2012) Experimental investigation of main controls to methane adsorption in clay-rich rocks. Appl Geochem 27:2533–2545

Jia Y, Song C, Wang J, Gan Q (2021) The breakdown process of low-permeable shale and high-permeable sandstone rocks due to non-aqueous fracturing: the role of fluid infiltration. J Nat Gas Sci Eng 89:103873

Jiang Y, Luo Y, Lu Y, Qin C, Liu H (2016) Effects of supercritical CO_2 treatment time, pressure, and temperature on microstructure of shale. Energy 97:173–181

Joubert JI, Grein CT, Bienstock D (1974) Effect of moisture on the methane capacity of American coals. Fuel 53:186–191

Kang SM, Fathi E, Ambrose RJ, Akkutlu IY, Sigal RF (2011) Carbon dioxide storage capacity of organic-rich shales. SPE J 16(04):842–855. https://doi.org/10.2118/134583-pa

Karayiğit AI, Mastalerz M, Oskay RG, Buzkan I (2018) Bituminous coal seams from underground mines in the Zonguldak Basin (NW Turkey): insights from mineralogy, coal petrography, rock-Eval pyrolysis, and meso- and microporosity. Int J Coal Geol 199:91–122

Katz BJ, Arango I (2018) Organic porosity: a geochemist's view of the current state of understanding. Org Geochem 123:1–16

Kaveh NS, Barnhoorn A, Wolf KH (2016) Wettability evaluation of silty shale caprocks for CO_2 storage. Int J Greenhouse Gas Control 49:425–435

Kelemen SR, Kwiatek LM (2009) Physical properties of selected block Argonne Premium bituminous coal related to CO_2, CH_4, and N_2 adsorption. Int J Coal Geol 77(1–2):2–9

Kerr RA (2010) Natural gas from shale bursts onto the scene. Science 328(5986):1624–1626

Khan JA, Padmanabhan E, Haq IU, Yekeen N, Mumtaz M, Prajapati S (2022) Experimental assessment of fracture initiation and wettability in low and high brittle shales by CO_2 foam fracturing fluids. Energy Fuels 36(15):8288–8300

Khosravi MH, Kheirollahi M, Liu B, Gentzis T, Liu K, Morta HB, Ostadhassan M (2023) Physico-chemo-mechanical impact of sc-CO_2 on shale formations: the Bakken. Gas Sci Eng 112:204945

Kim TH, Cho J, Lee KS (2017) Evaluation of CO_2 injection in shale gas reservoirs with multi-component transport and geomechanical effects. Appl Energy 190:1195–1206

Kim AG (1977) Estimating methane content of bituminous coalbeds from adsorption data. Report of Investigation 8245. U.S. Bureau of Mines, Washington DC

King GE (2010) Thirty years of gas shale fracturing: what have we learned? In: SPE annual technical conference and exhibition?. SPE, p 133456

Klaver J, Desbois G, Littke R, Urai JL (2015) BIB-SEM characterization of pore morphology and distribution in postmature to overmature samples from Haynesville and Bossier Shales. Mar Petrol Geol 59:451–466

Kolle JJ (2000) Coiled-tubing drilling with supercritical carbon dioxide. In: SPE/CIM international conference on horizontal well technology. SPE, p 65534

Krooss BM, van Bergen F, Gensterblum Y, Siemons N, Pagnier HJM, David P (2002a) High-pressure methane and carbon dioxide adsorption on dry and moisture-equilibrated Pennsylvanian coals. Int J Coal Geol 51:69–92

Krooss BV, Van Bergen F, Gensterblum Y, Siemons N, Pagnier HJM, David P (2002b) High-pressure methane and carbon dioxide adsorption on dry and moisture-equilibrated Pennsylvanian coals. Int J Coal Geol 51(2):69–92

Kuila U, Prasad M (2013) Specific surface area and pore-size distribution in clays and shales. In: Geophysical prospecting, 61(2-rock physics for reservoir exploration, characterisation and monitoring), pp 341–362

Kumar S, Chandra D, Hazra B, Vishal V, Pathegama Gamage R (2024) Nanopore Characteristics of Barakar Formation Shales and Their Impact on the Gas Storage Potential of Korba and Raniganj Basins in India. Energy & Fuels

Labani MM, Rezaee R, Saeedi A, Hinai AA (2013) Evaluation of pore size spectrum of gas shale reservoirs using low pressure nitrogen adsorption, gas expansion and mercury porosimetry: a case study from the Perth and Canning Basins, Western Australia. J Petrol Sci Eng 112:7–16

Lamberson MN, Bustin RM (1993) Coalbed methane characteristics of the Gates Formation coals, northeastern British Columbia: effect of maceral composition. Am Assoc Petrol Geol Bull 77:2062–2076

Lamond TG (1962) The sorption of gases on carbon: Ph.D. thesis, University of Durham, Durham

Larsen J (2004) The effects of dissolved CO_2 on coal structure and properties. Int J Coal Geol 57(1):63–70

Laxminarayana C, Crosdale PJ (1999) Role of coal type and rank on methane sorption characters of Bowen Basin, Australia coals. Int J Coal Geol 40:309–325

Lei X, Zhang S, Xu G, Zou Y (2015) Impact of perforation on hydraulic fracture initiation and extension in tight natural gas reservoirs. Energy Technol 3(6):618–624

Levine JR, Jonson PW, Beamish BB (1993) High-pressure microgravimetry provides a viable alternative to volumetric method in gas sorption studies on coal. In: Proceedings of the 1993 international coalbed methane symposium. University of Alabama, Tuscaloosa, Al, pp 187–195

Levy J, Day S, Killingley J (1997) Methane capacities of Bowen Basin coals related to coal properties. Fuel 76(9):813–819

Li Y, Zhang C, Tang D, Gan Q, Niu X, Wang K, Shen R (2017) Coal pore size distributions controlled by the coalification process: an experimental study of coals from the Junggar, Ordos and Qinshui basins in China. Fuel 206:352–363

Li W, Pang B, Su EL, Yang Q, Liu Q, Cheng Y (2019) Time-dependence of mechanical property alterations on anthracite coals treated by supercritical carbon dioxide. Geofluids 2019:1–9

Li G, Haizhu W, Zhonghou S, Song X, Guanwei T, Xinjuan H (2012) Experimental study on the efficiency of cuttings carrying with supercritical CO_2. In: IADC/SPE Asia pacific drilling technology conference and exhibition? SPE

Liu A, Liu S (2022) Mechanical property alterations across coal matrix due to water-CO_2 treatments: a micro-to-nano scale experimental study. Energy 248:123575

Liu B, Schieber J, Mastalerz M (2017) Combined SEM and reflected light petrography of organic matter in the New Albany Shale: a perspective on organic porosity development with thermal maturation. Int J Coal Geol 184:57–72

Liu B, Qi C, Mai T, Zhang J, Zhan K, Zhang Z, He J (2018) Competitive adsorption and diffusion of CH_4/CO_2 binary mixture within shale organic nanochannels. J Nat Gas Sci Eng 53:329–336

Liu S, Sang S, Ma J, Wang T, Du Y, Fang H (2019) Effects of supercritical CO_2 on micropores in bituminous and anthracite coal. Fuel 242:96–108

Liu S, Zhu C, Li Y, Hu S, Zhang X, Ma C (2023a) Mechanism of adsorption capacity enhancement of coal due to interaction with high-pressure $scCO_2$-water system. Gas Sci Eng 117:205080

Liu X, Jia X, Liu W, Nie B, Zhang C, Song D (2023b) Mechanical strength and porosity changes of bituminous coal induced by supercritical CO_2 interactions: influence of saturation pressure. Geoenergy Sci Eng 225:211691

Liu Y, Tang J, Jia Y, Lu Y, Shang C, Tian R, Sun X, Cheng Q, Zhao G (2023c) Investigation on capillary force recovery of nanoparticle-coated shale after supercritical CO_2 exposure: implications for CO_2 sequestration. Energy Fuels 37(17):13137–13150

Löhr SC, Baruch ET, Hall PA, Kennedy MJ (2015) Is organic pore development in gas shales influenced by the primary porosity and structure of thermally immature organic matter? Org Geochem 87:119–132

Long X, Liu Q, Ruan X, Kang Y, Lyu Q (2016) Numerical investigation of the flow of supercritical carbon dioxide injected into the bottom hole during drilling with special emphasis on the real gas effects. J Nat Gas Sci Eng 34:1044–1053

Loucks RG, Reed RM, Ruppel SC, Jarvie DM (2009) Morphology, genesis, and distribution of nanometer-scale pores in siliceous mudstones of the Mississippian Barnett Shale. J Sediment Res 79:848–861

Loucks RG, Reed RM, Ruppel SC, Hammes U (2012) Spectrum of pore types and networks in mudrocks and a descriptive classification for matrix-related mudrock pores. AAPG Bull 96:1071–1098

Lu XC, Li FC, Watson AT (1995) Adsorption measurements in Devonian shales. Fuel 74(4):599–603

Luo X, Ren X, Wang S (2019) Supercritical CO_2-water-shale interactions and their effects on element mobilization and shale pore structure during stimulation. Int J Coal Geol 202:109–127

Lyu Q, Long X, Ranjith PG, Kang Y (2016) Unconventional gas: experimental study of the influence of subcritical carbon dioxide on the mechanical properties of black shale. Energies 9(7):516

Lyu Q, Long X, Ranjith PG, Tan J, Kang Y, Wang Z (2018) Experimental investigation on the mechanical properties of a low-clay shale with different adsorption times in sub-/super-critical CO_2. Energy 147:1288–1298

Lyu Q, Wang K, Wanniarachchi WAM, Hu C, Shi J (2020) Hydro-mechanical properties of a low-clay shale with supercritical CO_2 imbibition. Geomech Geophys Geo-Energy Geo-Resour 6:1–14

Lyu Q, Tan J, Li L, Ju Y, Busch A, Wood DA, Ranjith PG, Middleton R, Shu B, Hu C, Wang Z (2021) The role of supercritical carbon dioxide for recovery of shale gas and sequestration in gas shale reservoirs. Energy Environ Sci 14(8):4203–4227

Ma Y, Yue C, Li S, Xu X, Niu Y (2018) Study of CH_4 and CO_2 competitive adsorption on shale in Yibin, Sichuan province of China. Carbon Resour Convers 11:5. https://doi.org/10.1016/j.crcon.2018.11.005

Mahajan OP (1984) Physical characterization of coal. Powder Technol 40:1–15

Mahajan OP (1991) CO_2 surface area of coals: the 25 year paradox. Carbon 29:735–742

Mahamud MM, Novo MF (2008) The use of fractal analysis in the textural characterization of coals. Fuel 87:222–231

Mann U, Neisel JD, Burchard WG, Heinen V, Welte DH (1994) Fluid-rock interfaces as revealed by cryo-scanning electron microscopy. First Break 12:131–136

Maranuk C, Rodriguez A, Trapasso J, Watson J (2014) Unique system for underbalanced drilling using air in the Marcellus Shale. In: SPE eastern regional meeting, pp 239–249

Mastalerz M, Gluskoter H, Rupp J (2004) Carbon dioxide and methane sorption in high volatile bituminous coals from Indiana, USA. Int J Coal Geol 60(1):43–55

Mastalerz M, Drobniak A, Schimmelmann A (2009) Changes in optical properties, chemistry, and micropore and mesopore characteristics of bituminous coal at the contact with dikes in the Illinois basin. Int J Coal Geol 77:310–319

Mastalerz M, Schimmelmann A, Lis GP, Drobniak A, Stankiewicz A (2012) Influence of maceral composition on geochemical characteristics of immature shale kerogen: insight from density fraction analysis. Int J Coal Geol 103:60–69

Mastalerz M, Schimmelmann A, Drobniak A, Chen Y (2013) Porosity of Devonian and Mississippian New Albany Shale across a maturation gradient: insights from organic petrology, gas adsorption, and mercury intrusion. AAPG Bull 97:1621–1643

Medek J, Weishauptová Z, Kovář L (2006) Combined isotherm of adsorption and absorption on coal and differentiation of both processes. Micropor Mesopor Mater 89(1–3):276–283

Meng M, Qiu Z (2018) Experiment study of mechanical properties and microstructures of bituminous coals influenced by supercritical carbon dioxide. Fuel 219:223–238
Meng S, Jin X, Tao J, Wang X, Zhang C (2021) Evolution characteristics of mechanical properties under supercritical carbon dioxide treatment in shale reservoirs. ACS Omega 6(4):2813–2823
Middleton RS, Carey JW, Currier RP, Hyman JD, Kang Q, Karra S, Jiménez-Martínez J, Porter ML, Viswanathan HS (2015) Shale gas and non-aqueous fracturing fluids: opportunities and challenges for supercritical CO_2. Appl Energy 147:500–509
Milewska-Duda J, Duda J, Nodzeñski A, Lakatos J (2000) Absorption and adsorption of methane and carbon dioxide in hard coal and active carbon. Langmuir 16:5458–5466
Milliken KL, Rudnicki M, Awwiller DN, Zhang T (2013) Organic matter-hosted pore system, Marcellus Formation (Devonian), Pennsylvania. AAPG Bull 97:177–200
Mishra DK, Samad SK, Varma AK, Mendhe VA (2018) Pore geometrical complexity and fractal facets of Permian shales and coals from Auranga Basin, Jharkhand, India. J Nat Gas Sci Eng 52:25–43
Modica CJ, Lapierre SG (2012) Estimation of kerogen porosity in source rocks as a function of thermal transformation: example from the Mowry Shale in the Powder River Basin of Wyoming. AAPG Bull 96:87–108
Moore TA (2012) Coalbed methane: a review. Int J Coal Geol 101:36–81
Mukherjee M, Misra S (2018) A review of experimental research on enhanced coal bed methane (ECBM) recovery via CO_2 sequestration. Earth Sci Rev 179:392–410
Murray DK (1991) Coal bed methane; natural gas resources from coal seams. In: Peters DC (ed) Geology in coal resource utilization. Tech Books, United States, pp 97–103
Nuttal BC, Eble C, Bustin RM, Drahovzal JA (2005) Analysis of Devonian black shales in Kentucky for potential carbon dioxide sequestration and enhanced natural gas production. Greenhouse gas control technologies, vol 7. Elsevier, Amsterdam, pp 2225–2228
Nuttal BC, Eble C, Bustin RM, Drahovzal JA (2003) Analysis of Devonian black shales in Kentucky for potential carbon dioxide sequestration and enhanced natural gas production. In: UNT Libraries Government Documents Department. Digital
O'Gorman JV (1971) Ph.D Thesis. The Pennsylwutia State University
Okolo GN, Everson RC, Neomagus HWJP, Roberts MJ, Sakurovs R (2015) Comparing the porosity and surface areas of coal as measured by gas adsorption, mercury intrusion and SAXS techniques. Fuel 141:293–304
Ougier-Simonin A, Renard F, Boehm C, Vidal-Gilbert S (2016) Microfracturing and microporosity in shales. Earth Sci Rev 162:198–226
Ozotta O, Liu K, Gentzis T, Carvajal-Ortiz H, Liu B, Rafieepour S, Ostadhassan M (2021) Pore structure alteration of organic-rich shale with Sc-CO_2 exposure: the Bakken formation. Energy Fuels 35(6):5074–5089
Pan Y, Hui D, Luo P, Zhang Y, Sun L, Wang K (2018) Experimental investigation of the geochemical interactions between supercritical CO_2 and shale: implications for CO_2 storage in gas-bearing shale formations. Energy Fuels 32(2):1963–1978
Parkash S, Chakrabartty SK (1986) Microporosity in Alberta plains coals. Int J Coal Geol 6:55–70
Perera MSA (2018) A comprehensive overview of CO_2 flow behaviour in deep coal seams. Energies 11(4):906
Perera MSA, Ranjith PG, Viete DR (2013) Effects of gaseous and super-critical carbon dioxide saturation on the mechanical properties of bituminous coal from the Southern Sydney Basin. Appl Energy 110:73–81
Pommer M, Milliken K (2015) Pore types and pore-size distributions across thermal maturity, Eagle Ford Formation, southern Texas. AAPG Bull 99:1713–1744
Pozo M, Pino D, Bessieres D (2017) Effect of thermal events on maturation and methane adsorption of Silurian black shales (Checa, Spain). Appl Clay Sci 136:208–218. https://doi.org/10.1016/j.clay.2016.11.026
Qi R, Ning Z, Wang Q, Huang L, Wu X, Cheng Z, Zhang W (2019) Measurements and modeling of high-pressure adsorption of CH_4 and CO_2 on shales. Fuel 242:728–743

Qin C, Jiang Y, Luo Y, Xian X, Liu H, Li Y (2017) Effect of supercritical carbon dioxide treatment time, pressure, and temperature on shale water wettability. Energy Fuels 31(1):493–503

Qin C, Jiang Y, Kang Z, Song X, Liu H (2021) Experimental study on tensile strength and acoustic emission characteristics of shale exposure to supercritical CO_2. Energy Sour A Recov Utiliz Environ Effects 43(8):977–992

Qutob H, Byrne M (2015) Formation damage in tight gas reservoirs. In: SPE European formation damage conference and exhibition. SPE

Radovic LR, Menon VC, Leon Y, Leon CA, Kyotani T, Danner RP, Anderson S, Hatcher PG (1997) On the porous structure of coals: evidence for an interconnected but constricted micropore system and implications for coalbed methane recovery. Adsorption 3:221–232

Reucroft PJ, Patel KB (1983) Surface area and swellability of coal. Fuel 62(3):279–284

Rice DD, Law BE, Clayton JL (1993) Coalbed gas—an undeveloped resource. US Geol Surv Profess Paper 1570:389–404

Rogelj J, Popp A, Calvin KV, Luderer G, Emmerling J, Gernaat D, Fujimori S, Strefler J, Hasegawa T, Marangoni G, Krey V (2018) Scenarios towards limiting global mean temperature increase below 1.5 C. Nat Clim Change 8(4):325–332

Rogers R (1994) Coalbed methane: principles and practices. The Prentice Hall Professional, Technical, and Reference Division

Ross DJ, Bustin RM (2009) The importance of shale composition and pore structure upon gas storage potential of shale gas reservoirs. Mar Petrol Geol 26(6):916–927

Rouquerol J, Avnir D, Fairbridge CW, Everett DH, Haynes JH, Pernicone N, Ramsay JDF, Sing KSW, Unger K (1994) Recommendations for the characterization of porous solids, international union of pure and applied chemistry. Pure Appl Chem 68:1739–1758

Ryan B (1992) An equation for estimation of maximum coalbed-methane resource potential: British Columbia geological survey. Geological Fieldwork 1991, Paper 1992-1, pp 393–395

Safaei-Farouji M, Misch D, Sachsenhofer RF (2023) A review of influencing factors and study methods of carbon capture and storage (CCS) potential in coals. Int J Coal Geol 12:104351

Sakhaee-Pour A, Bryant SL (2012) Gas permeability of shale. SPE Reserv Evaluat Eng 15(04):401–409

Sanguinito S, Goodman A, Tkach M, Kutchko B, Culp J, Natesakhawat S, Fazio J, Fukai I, Crandall D (2018) Quantifying dry supercritical CO_2-induced changes of the Utica Shale. Fuel 226:54–64

Schettler PD, Parmely CR, Juniata C (1991) Contributions to total storage capacity in Devonian shales. SPE paper, 23422

Schieber J, Lazar R, Bohacs K, Klimentidis B, Ottmann J, Dumitrescu M (2016) An SEM study of porosity in the Eagle Ford Shale of Texas: pore types and porosity distribution in a depositional and sequence stratigraphic context. AAPG Mem 110:153–172

Schnoor JL (2012) Shale gas and hydrofracturing. Environ Sci Technol 46(9):4686–4686

Schoeling L, McGovern M (2000) Pilot test demonstrates how CO_2 injection enhances coalbed methane recovery. Petrol Technol Dig A World Oil Mag 14:9

Shahbazi K, Mehta SA, Moore RG, Ursenbach MG, Fraassen KCV (2007) Investigation of explosion occurrence in underbalanced drilling. In: SPE Oklahoma City oil and gas symposium/production and operations symposium. SPE

Shen Z, Wang H, Li G (2011) Numerical simulation of the cutting-carrying ability of supercritical carbon dioxide drilling at horizontal section. Petrol Explor Develop 38(2):233–236

Shen Y, Pang Y, Shen Z, Tian Y, Ge H (2018) Multiparameter analysis of gas transport phenomena in shale gas reservoirs: apparent permeability characterization. Nat Sci Rep 8(2601):1–14. https://doi.org/10.1038/s41598-018-20949

Shi X, Jiang S, Wang Z, Bai B, Xiao D, Tang M (2020) Application of nanoindentation technology for characterizing the mechanical properties of shale before and after supercritical CO_2 fluid treatment. J CO2 Utiliz 37:158–172

Smith IM (1999) CO_2 reduction-prospects for coal. IEA Coal Research, CCC/26. IEA Coal Research, London, p 84

Sondergeld CH, Ambrose RJ, Rai CS, Moncrieff J (2010) Micro-structural studies of gas shales. In SPE unconventional gas conference, Feb. 23–25, Pittsburgh, PA, Paper No. SPE-131771. OnePetro

Song W, Ni H, Tang P, Zhang S, Gao J, Zhang J, Shen B (2021) Simulation of supercritical carbon dioxide fracturing in shale gas reservoir. J Therm Sci 30:1444–1451

Su E, Liang Y, Chang X, Zou Q, Xu M, Sasmito AP (2020) Effects of cyclic saturation of supercritical CO_2 on the pore structures and mechanical properties of bituminous coal: an experimental study. J CO2 Utiliz 40:101208

Sun W, Feng Y, Jiang C, Chu W (2015) Fractal characterization and methane adsorption features of coal particles taken from shallow and deep coalmine layers. Fuel 155:7–13

Tan J, Weniger P, Krooss B, Merkel A, Horsfield B, Zhang J, Boreham CJ, van Graas G, Tocher BA (2014) Shale gas potential of the major marine shale formations in the Upper Yangtze Platform, South China, Part II: Methane sorption capacity. Fuel 129:204–218

Tao Z, Bielicki JM, Clarens AF (2014) Physicochemical factors impacting CO_2 sequestration in depleted shale formations: the case of the Utica shale. Energy Proced 63:5153–5163. https://doi.org/10.1016/j.egypro.2014.11.545

Thomas KM (2023) Perspectives of gas adsorption and storage in Kerogens and Shales. Energy Fuels 37(4):2569–2585

Thomas J, Damberger HH (1976) Internal surface area, moisture content, and porosity of Illinois coals: variations with coal rank. Circular no. 493

Tian S, He Z, Li G, Wang H, Shen Z, Liu Q (2016) Influences of ambient pressure and nozzle-to-target distance on SC-CO_2 jet impingement and perforation. J Nat Gas Sci Eng 29:232–242

U.S. Energy Information Administration (EIA). Annual energy outlook 2016 with projections to 2040. https://www.eia.gov/outlooks/aeo/pdf/0383(2016).pdf

Unsworth JF, Fowler CS, Jones LF (1989) Moisture in coal: 2. Maceral effects on pore structure. Fuel 68(1):18–26

Valenza JJ, Drenzek N, Marques F, Pagels M, Mastalerz M (2013) Geochemical controls on shale microstructure. Geology 41:611–614

Varma AK, Hazra B, Samad SK, Panda S, Mendhe VA (2014) Methane sorption dynamics and hydrocarbon generation of shale samples from West Bokaro and Raniganj basins, India. J Nat Gas Sci Eng 21:1138–1147

Varma AK, Biswal S, Hazra B, Mendhe VA, Misra S, Samad SK, Singh BD, Dayal AM, Mani D (2015) Petrographic characteristics and methane sorption dynamics of coal and shaly-coal samples from Ib Valley Basin, Odisha, India. Int J Coal Geol 141:51–62

Veytskin YB, Tammina VK, Bobko CP, Hartley PG, Clennell MB, Dewhurst DN, Dagastine RR (2017) Micromechanical characterization of shales through nanoindentation and energy dispersive x-ray spectrometry. Geomech Energy Environ 9:21–35

Vishal V (2017) Saturation time dependency of liquid and supercritical CO_2 permeability of bituminous coals: implications for carbon storage. Fuel 192:201–207

Vishal V, Chandra D, Bahadur J, Sen D, Hazra B, Mahanta B, Mani D (2019) Interpreting pore dimensions in gas shales using a combination of SEM imaging, small-angle neutron scattering, and low-pressure gas adsorption. Energy Fuels 33(6):4835–4848

Vishal V, Chandra D, Singh U, Verma Y (2021a) Understanding initial opportunities and key challenges for CCUS deployment in India at scale. Resour Conserv Recycl 175:105829

Vishal V, Verma Y, Chandra D, Ashok D (2021b) A systematic capacity assessment and classification of geologic CO_2 storage systems in India. Int J Greenhouse Gas Control 111:103458

Walker PL, Verma SK, Rivera-Utrilla J, Khan R (1988) A direct measurement of expansion in coals and macerals induced by carbon dioxide and methanol. Fuel 67:719–726

Wan T, Mu Z (2018) The use of numerical simulation to investigate the enhanced Eagle Ford shale gas condensate well recovery using cyclic CO_2 injection method with nano-pore effect. Fuel 233:123–132

Wang J, Elsworth D (2020) Fracture penetration and proppant transport in gas-and foam-fracturing. J Nat Gas Sci Eng 77:103269

Wang H, Shen Z, Li G (2012) The development and prospect of supercritical carbon dioxide drilling. Petrol Sci Technol 30(16):1670–1676

Wang S, Song Z, Cao T, Song X (2013a) The methane sorption capacity of Paleozoic shales from the Sichuan Basin, China. Mar Pet Geol 44:112–119

Wang S, Song Z, Cao T, Song X (2013b) The methane sorption capacity of Paleozoic shales from the Sichuan Basin, China. Mar Pet Geol 44:112–119. https://doi.org/10.1016/j.marpetgeo.2013.03.007

Wang L, Yao B, Xie H, Winterfeld PH, Kneafsey TJ, Yin X, Wu YS (2017) CO_2 injection-induced fracturing in naturally fractured shale rocks. Energy 139:1094–1110

Wang H, Li G, He Z, Tian S, Wang M, Yang B, Lu Q, Weng L (2018a) Experimental investigation on abrasive supercritical CO_2 jet perforation. J CO2 Utiliz 28:59–65

Wang J, Elsworth D, Wu Y, Liu J, Zhu W, Liu Y (2018b) The influence of fracturing fluids on fracturing processes: a comparison between water, oil and SC-CO_2. Rock Mech Rock Eng 51:299–313

Wang S, Liu K, Han J, Ling K, Wang H, Jia B (2019) Investigation of properties alternation during super-critical CO_2 injection in shale. Appl Sci 9(8):1686

Wang F, Zhang X, Wu C, Zhang S, Wang K (2023) Mechanism of supercritical CO_2 on the chemical structure and composition of high-rank coals with different damage degrees. Fuel 344:128027

Wang W, Wang K, Zhang X, Jiang Y, Cai T, Yan J, Yue S (2024) Mechanical properties of coal-rock combined bodies under the action of ScCO_2: a study of damage and failure mechanism. Eng Fail Anal 156:107807

Wang FP, Reed RM (2009) Pore networks and fluid flow in gas shales: SPE Annual technical conference and exhibition, 4–7 October 2009. New Orleans, Louisiana: SPE Document ID

Wang C, Liu B, Mohammadi MR, Fu L, Fattahi E, Motra HB, Hazra B, Hemmati-Sarapardeh A, Ostadhassan M (2024) Integrating experimental study and intelligent modeling of pore evolution in the Bakken during simulated thermal progression for CO_2 storage goals. Appl Energy 359:122693

Weishauptova Z, Pribyl O, Sykorova I, Machovic V (2015) Effect of bituminous coal properties on carbon dioxide and methane high pressure sorption. Fuel 139:115–124

Weniger P, Kalkreuth W, Busch A, Krooss BM (2010) High-pressure methane and carbon dioxide sorption on coal and shale samples from the Paraná Basin. Brazil. J Int J Coal Geol 84(3–4):190–205. https://doi.org/10.1016/j.coal.2010.08.003

Weniger P, Francu J, Hemza P, Krooss BM (2012) Investigations on the methane and carbon dioxide sorption capacity of coals from the SW Upper Silesian Coal Basin, Czech Republic. Int J Coal Geol 93:23–39

Whiticar MJ (1994) Correlation of natural gases with their sources. In: Magoon LB, Dow WG (eds) The petroleum system—from source to trap: memoir, 60. AAPG, Tulsa, pp 261–283

Wood DA, Hazra B (2017) Characterization of organic-rich shales for petroleum exploration and exploitation: a review—part 1: bulk properties, multi-scale geometry and gas adsorption. J Earth Sci 28(5):739–757

Xie J, Cheng W, Wang R, Jiang G, Sun D, Sun J (2018) Experiments and analysis on the influence of perforation mode on hydraulic fracture geometry in shale formation. J Petrol Sci Eng 168:133–147

Xie W, Chen S, Wang M, Yu Z, Wang H (2021) Progress and prospects of supercritical CO_2 application in the exploitation of shale gas reservoirs. Energy Fuels 35(22):18370–18384

Xu H, Zhu Y, Liu H, Ding H, Fang H, Zhang K, Jia J (2023) Wettability alteration and dynamic wetting behavior of coal during geologic CO_2 sequestration using LF-NMR technology. Fuel 354:129355

Xu H, Hu J, Liu H, Ding H, Zhang K, Jia J, Fang H, Gou B (2024) Effect of the interaction time of CO_2–H_2O on the alterations of coal pore morphologies and water migration during wetting. Energy 294:130944

Yang K, Zhou J, Xian X, Zhang C, Gan Q, Dong Z (2023a) Effect of supercritical CO_2-water-shale interaction on mechanical properties of shale and its implication for carbon sequestration. Gas Sci Eng 111:204930

Yang Y, Xing P, Dai L, Liu X, Nie B (2023b) Coal microstructural and mechanical alterations induced by supercritical CO_2 exposure: role of water. Fuel 352:128952

Yang K, Zhou J, Xian X, Zhou L, Zhang C, Lu Z, Yin H (2024) Changes of wettability of shale exposed to supercritical CO_2-water and its alteration mechanism: implication for CO_2 geo-sequestration. Fuel 357:129942

Yao Y, Liu D, Tang D, Tang S, Huang W (2008) Fractal characterisation of adsorption-pores of coals from North China: an investigation on CH_4 adsorption capacity of coals. Int J Coal Geol 73:27–42

Yee D, Seidle JP, Hanson WB (1993) Gas sorption on coal and measurement of gas content. In: Law BE, Rice DD (eds) Hydrocarbons from coal. American Association of Petroleum Geologists, Studies in Geology 38, Tulsa, Oklahoma, Studies in Geology 38, pp 203–218

Zhang X, Lu Y, Tang J, Zhou Z, Liao Y (2017) Experimental study on fracture initiation and propagation in shale using supercritical carbon dioxide fracturing. Fuel 190:370–378

Zhang Y, He J, Li X, Lin C (2019) Experimental study on the supercritical CO_2 fracturing of shale considering anisotropic effects. J Pet Sci Eng 173:932−940

Zhang CP, Liu S, Ma ZY, Ranjith PG (2021) Combined micro-proppant and supercritical carbon dioxide (SC-CO_2) fracturing in shale gas reservoirs: a review. Fuel 305:121431

Zhang T, Ellis GS, Ruppel SC, Milliken K, Yang R (2012) Effect of organic-matter type and thermal maturity on methane adsorption in shale-gas systems 47:120–131

Zhou D, Zhang G, Zhao P, Wang Y, Xu S (2018) Effects of post-instability induced by supercritical CO_2 phase change on fracture dynamic propagation. J Petrol Sci Eng 162:358–366

Zhou D, Zhang G, Prasad M, Wang P (2019a) The effects of temperature on supercritical CO_2 induced fracture: an experimental study. Fuel 247:126–134

Zhou J, Liu M, Xian X, Jiang Y, Liu Q, Wang X (2019b) Measurements and modelling of CH_4 and CO_2 adsorption behaviors on shales: implication for CO_2 enhanced shale gas recovery. Fuel 251:293–306

Zhu C, Wan J, Tokunaga TK, Liu N, Lin B, Wu H (2019) Impact of CO_2 injection on wettability of coal at elevated pressure and temperature. Int J Greenhouse Gas Control 91:102840

Zhu Z, Li M, Lin M, Peng B, Sun L, Chen L (2011) Investigation on variations in wettability of reservoir rock induced by CO_2-brine-rock interactions. In: SPE EUROPEC/EAGE annual conference and exhibition. OnePetro

Zou Y, Li N, Ma X, Zhang S, Li S (2018) Experimental study on the growth behavior of supercritical CO_2-induced fractures in a layered tight sandstone formation. J Nat Gas Sci Eng 49:145–156

Chapter 7
Conclusions

For thorough characterization of a CBM or gas shale reservoir, and to assess its gas recovery and/or CO_2 storage potential, a detailed assessment of its organic and mineral matter characteristics, sorption properties, and flow attributes through its pore-fracture system is of paramount importance. Generating reliable geochemical data in the laboratory is one of the first requisites for prospect or play appraisals. While analysing subsurface shales and coals, often the Rock–Eval technique is used, as the data provides important insights viz. thermal maturity stage, kerogen type, organic richness, and free hydrocarbons present within the samples. Although the overall process is simple and the results produce are easily interpreted, different factors influence the results. Particle crush-size and amount of sample used for analysis affect the results, and analysts should cross-check the nature of the resultant curves generated from analysis for improving the accuracy of the dataset generated. FID detector response (saturated or undersaturated or reliable), nature of S2 curves (asymmetrical or Gaussian-shaped or spiky), nature of S3 (whether CO_2 during pyrolysis is generated from decomposition of organic matter or if there's any influence of carbonate minerals) and S4 (whether entire area is calculated for TOC; also, especially significant for low-TOC or overmatured samples) curves, all provide critical insights about the unconventional play being targeted. Thorough and careful consideration of the above-mentioned factors, not only allows generation of precise data, but helps is predicting the nature of the sample being studied.

For the technical development of a CBM well, or shale gas reservoir, and for CO_2-ECBM and CO_2-EGR as well, the role of the pore systems is extremely significant along with detailed understanding of fluid transport and gas sorption mechanism. Clarity on the mechanisms of adsorption, adsorption models, the techniques used for laboratory high-pressure and low-pressure gas adsorption studies all are extremely critical for gas-in-place estimates and subsurface CO_2 sequestration. It has frequently been reported that organic matter type and its thermal maturity primarily controls gas storage capacity of coals and shales, although mineral matter is also known to play an important role in storage and flow attributes of shales. However, its over

B. Hazra et al., *Unconventional Hydrocarbon Reservoirs: Coal and Shale*, Petroleum Engineering, https://doi.org/10.1007/978-3-031-53484-3_7

simplistic to consider that only organic matter type and thermal maturity controls the adsorption potential of these reservoirs, especially CBM reservoirs. Additional factors viz. moisture content, temperature etc., can also significantly affect the results. Competitive adsorption between CH_4 and CO_2, and preferential adsorption of the later relative to the former gas, is especially significant for CO_2-ECBM and CO_2-EGR. Results from different researchers clearly demonstrates that organic matter in coals and shales have several times higher affinity towards CO_2, than CH_4. This allows injecting CO_2 into producing wells at appropriate stages to sweep off additional CH_4, and permanently sequester the injected CO_2.

It is imperative to consider that the CO_2, when injected in deep shale or coal reservoirs, undergo phase transition to supercritical state beyond its critical point, and is known as supercritical CO_2 (SC-CO_2). SC-CO_2 facilitates CO_2 sequestration and also boosts drilling efficiency in unconventional reservoir development. As drilling fluid, SC-CO_2 mitigates overbalance conditions and offers efficient downhole motor operation. It boasts advantages such as high penetration rates, low frictional head loss, and enhanced recovery through fracturing. However, challenges include compatibility issues with elastomers, which may swell, and the need to address solids transportation concerns. Despite these challenges, SC-CO_2 drilling shows great potential, with ongoing research focusing on optimizing its application in the oil and gas industry. Additionally, SC-CO_2 offers advantages in unconventional rock fracturing over aqueous fluids, yielding complex fracture networks with reduced formation damage. SC-CO_2 induces irregular cracks and complex fracture networks, outperforming hydraulic fracturing in fractured rock volume. While it surpasses water-based fluids in stimulated reservoir volume, challenges include proppant carrying limitations. Microproppants improve proppant distribution but face cost and safety concerns. Perforation angle affects breakdown pressure, and temperature impacts the effectiveness of SC-CO_2, requiring further field evaluation. Studies reveal decreased elastic modulus and strength in shales post SC-CO_2 saturation, influenced by factors such as pressure, temperature, and exposure duration. Alterations in mineral, pore structure, and surface wettability are observed, emphasizing the need for a nuanced understanding to optimize processes in shale reservoirs. The intricate interactions between SC-CO_2 and shale underscore the importance of comprehensive research for advancing hydrocarbon extraction and carbon capture and sequestration strategies in shale reservoirs. Furthermore, the injection of SC-CO_2, significantly impacts the mechanical properties and structural integrity of coal. SC-CO_2 induces greater strength reduction compared to gaseous CO_2 and N_2, posing potential risks such as premature breakthrough of CO_2 from coal seam gas production wells. Understanding these complexities is vital for optimizing CO_2 injection strategies while minimizing risks and ensuring effective gas storage and recovery.

As gas in unconventional coal and shale reservoirs primarily exists in adsorbed state and as pores are the primary gas storage sites in these reservoirs, understanding the pores and their connectivity is of utmost importance. Low pressure gas adsorption has been observed to be extremely effective in evaluating the pore structural facets of coals and shales. While the data generated from low pressure gas adsorption–desorption experiments is extremely critical and provides valuable insights, the results can be largely influenced by the analytical protocols followed. Impact of particle crush size, degassing temperature and time can impact the results produced, depending upon the type of kerogen present within the samples. Some experimental limitations are also noted while examining complex pore structures in organic-rich horizons using N_2 as the adsorbent, and CO_2 over N_2, has been observed to be more useful evaluating the micropore characteristics of such samples.

In addition to gas adsorption, imaging techniques have also been employed to evaluate the microstructures of shales and coals. Balancing imaging resolution and field of view (FOV) is challenging in geological imaging techniques. Increasing FOV brings the scan area closer to the representative elementary volume (REV) but sacrifices resolution, limiting the ability to capture both macroscopic and microscopic details in heterogeneous shale and coal. Additionally, for accurate imaging, it is crucial to follow a meticulous sample preparation process. Due to the heterogenous nature of coal and shale utilizing different sample sections for distinct imaging approaches could yield different inferences. Most common 2D imaging techniques are optical microscopy, SEM and TEM. While optical microscopy is essential for obtaining key information on the organic matter (type and composition) and mineral matrix distribution it is limited by its low resolution compared to SEM and TEM that can achieve resolution of up to 2 nm. In transitioning from 2 to 3D imaging, FIB-SEM stands out, operating like SEM but with in-situ ion-milling for layer-by-layer scanning, enabling depth information up to a few millimetres. While advantageous for creating 3D volumes of mesopores in organic matter, its smaller scan volume limits its ability to quantify the heterogeneity and anisotropy in shale and coal. Image processing is crucial for quantifying geological samples, employing filters and segmentation techniques. Thresholding, ML, and watershed segmentation enhance accuracy in feature extraction from 2 and 3D coal and shale images. Research on in-situ time-lapse imaging should be given more focus, which will reveal a new dimension in understanding the flow/deformation/thermal evolution of coal and shale.

The optical-electron correlative microscopy technique employed on shales based on optimal contrast allows distinguishing the DOM and mineral matter at acceleration voltages of 15 and 10 kV under BSE and SE2 scanning modes, respectively. Additionally, selection of right working distance during electron microscopy is crucial. Vitrinite is difficult to identify under SEM alone without the correlative technique. On the other hand, certain inertinite macerals like fusinite or semifusinite could be recognized through SEM when cell lumens are present; nonetheless, without the correlative technique, identification might pose challenges when distinctive diagnostic features are lacking. Liptinite macerals, exhibiting a dark gray to black color and distinctive morphology, such as sporinite, are relatively easy to differentiate under SEM. The correlative technique facilitates the detection of SEM-visible pores within organic matter in shale/coal samples, providing a means to establish a correlation between the observed porosity and the maceral type. Identifying the association between mineral matter and organic content in shale/coal poses challenges with an optical microscope alone. The correlative technique proves instrumental, enabling the identification of minerals associated with organic matter under SEM through EDS analysis. This approach significantly enhances the precision in the characterization of organic and mineral components in shale/coal. It should be noted, however, that the correlative technique has certain limitations, notably its inability to detect pores smaller than 2 nm which could result in inaccurate assessment of porosity.

Upscaling, achieved through systematic averaging and minimizing heterogeneities, accelerates simulations without significant accuracy loss. Balancing detail loss and efficiency gain has become a key focus of researchers. Multimodal porosity and permeability models are used for modelling fluid flow in complex porous system like shale and coal, where the physics of flow in fractures and matrix are completely different. Capturing transient flow and phase segregation in matrix blocks requires subdividing the matrix into smaller subdomains, although this increases computational time. Permeability and flow behavior within the matrix play a significant role, leading to the development of discrete fracture network (DFN) models. Modern dual/triple porosity and permeability models demonstrate better history matching for shale gas and coal bed methane (CBM) simulations. Several models have been suggested to tackle the intricate fluid transport complexities at different scales in shale, considering parameters such as pore radius, Knudsen number, and diffusion coefficient. The advent of apparent permeability models has enhanced the realism of simulating flow through nanopores, incorporating considerations for the wettability of the media. Conventional reservoir models neglect the coupling of geomechanics and transport phenomena, underscoring the necessity for integrated models to ensure precise simulations. Utilizing different extents of coupling methods has been instrumental, the most advanced form being fully coupled model, which is capable of solving geomechanics and flow simultaneously. Incorporating geomechanics in simulations markedly impacts gas and water recovery rates, underscoring the pivotal role of integrated modelling in optimizing hydrocarbon extraction strategies. Machine learning (ML) and artificial intelligence (AI) have transformed geological and geophysical data interpretation, enhanced real-time predictions and handling massive datasets. ML, including techniques like Artificial Neural Networks (ANN), Fuzzy Inference Systems, Genetic Algorithms, Support Vector Machines, and Decision Trees, facilitates accurate petrophysical and geomechanical analyses. These models learn from training data, improving over time with minimal human intervention. They address challenges like gap filling in seismic or log data, making them valuable tools in reservoir analysis. Numerous applications include generating synthetic geomechanical logs, brittleness index calculation, and production forecasting. ML/AI models contribute to optimizing well placement, enhancing recovery predictions, and streamlining reservoir management for unconventional resources, significantly impacting the oil and gas industry. Various approaches exist for upscaling small-scale experimental/logging data to reservoir-scale models, with simulation codes primarily tailored for conventional oil and gas operations but adapted for unconventional resources. Commercial simulators like ECLIPSE, PETREL, and INTERSECT by SLB, along with CMG-GEM and COMET3, are well-documented and widely used. TOUGH, SIMED, and FLAC are powerful tools, which can leverage coupling of multiphase flow with geomechanics. Sequential coupling is evident in GEOSIM and RocFlow, while fully coupled methods are emerging. Ongoing advancements include GPU acceleration, ML algorithms, and code parallelization for more accurate and efficient simulations.